# USING ANIMALS
# FOR CLOTHING

# Titles in the series:

**Using Animals
for Food**

**Using Animals
for Research**

**Using Animals
for Entertainment**

**Using Animals
for Clothing**

EXPLORING ANIMAL RIGHTS AND ANIMAL WELFARE

VOLUME 4

# USING ANIMALS FOR CLOTHING

GREENWOOD PRESS

Westport, Connecticut • London

**Library of Congress Cataloging-in-Publication Data**

Creative Media Applications
Trumbauer, Lisa, 1963–
    Exploring animal rights and animal welfare / Lisa Trumbauer.
      v. cm.
    Contents: v. 1. Using animals for food—v. 2. Using animals for research—v. 3. Using animals for entertainment—v. 4. Using animals for clothing.
    Includes bibliographical references and index.
    ISBN 0-313-32245-7 (set)—ISBN 0-313-32246-5 (v. 1)—ISBN 0-313-32247-3 (v. 2)—ISBN 0-313-32248-1 (v. 3)—ISBN 0-313-32249-X (v. 4)
      1. Animal welfare—Juvenile literature. 2. Animal rights—Juvenile literature. 3. Human-animal relationships—Juvenile literature. 4. Humane education—Juvenile literature. [1. Animals—Treatment. 2. Animal rights. 3. Human-animal relationships.] I. Title. II. Series.

HV4712.T78 2002
179.3—dc21                                           2002075303

British Library Cataloguing in Publication Data is available.

Library of Congress Catalog Card Number: 2002075303
ISBN: 0-313-32245-7 (set)
      0-313-32246-5 (Vol. 1)
      0-313-32247-3 (Vol. 2)
      0-313-32248-1 (Vol. 3)
      0-313-32249-X (Vol. 4)

First published in 2002

Greenwood Press, 88 Post Road West, Westport, CT 06881
An imprint of Greenwood Publishing Group, Inc.
www.greenwood.com

Printed in the United States of America

The paper used in this book complies with the Permanent Paper Standard issued by the National Information Standards Organization (Z39.48–1984)

10  9  8  7  6  5  4  3  2  1

*Photo Credits:*
Cover Image: AP/Wide World Photographs
AP/Wide World Photographs: Pages 6, 12, 14, 19, 20, 23, 28, 31, 32, 39, 40, 48, 51, 53, 58, 61, 63, 72, 74, 78, 80, 84, 88, 91, 92, 95, 98, 99, 100, 102, 104, 109, 113
©Gianni Dagli Orti/CORBIS: Page 2
©Historical Pictures Archives/CORBIS: Page 68

A Creative Media Applications, Inc. Production

Writer: Lisa Trumbauer
Design and Production: Alan Barnett, Inc.
Editor: Matt Levine
Copyeditor: Laurie Lieb
Proofreader: Betty C. Pessagno
AP Photo Researcher: Yvette Reyes
Consultants: Sheryl Dickstein-Pipe, Ph.D. and
    Stephen Zawistowski, Ph.D., Certified Applied Animal Behaviorist

# EXPLORING ANIMAL RIGHTS AND ANIMAL WELFARE

## VOLUME 4

# USING ANIMALS FOR CLOTHING

CONTENTS

# Introduction

Here's a riddle for you. What is something that you see every day that everyone has, including yourself? The answer: clothing!

Humans are the only animals on the planet who feel the necessity to wear clothing. In fact, humans have been wearing some form of clothing for thousands of years.

But why did humans first start to wear clothing? And what were the materials from which this clothing was made? Let's take a quick look back at some very ancient history.

## THE BEGINNINGS OF CLOTHING

Scientists believe that the early human *Homo sapiens,* or *Cro-Magnon,* first appeared about 30,000 to 40,000 years ago. Cro-Magnon people, the most modern of early humans, were more advanced than other mammals that inhabited Earth. Not only did they hunt, fish, and build shelters, but they also made tools and clothing.

Clothing served two purposes for *Homo sapiens.* One purpose was for warmth. As *Homo sapiens* traveled from Africa into the colder regions of Europe and Asia, and eventually across a land bridge into North America, the humans' bodies needed protection from the cold. Animal skins were good protection and were used for clothing. At first, the skins were simply draped or wrapped around the body. Later, the skins were sewn together to fit the body more precisely. Sewing was often a job

1

*This print from the 1800s shows a warrior or hunter wearing animal-skin clothing.*

done by the women. The women would sew together sturdy clothes for the men to wear while hunting. Clothing worn by women was typically simpler than the men's clothing and therefore did not require such sturdy construction.

The second purpose for which *Homo sapiens* wore animal skins is similar to the reason why some people wear clothing today: to show off! A person who wore the skin of a lion had to kill the animal first, which was not an easy task. Wearing the lion's skin showed how strong and brave that person was. Likewise, children who wore animal skins showed that their parents were good providers.

Historically, the next material most likely used to make clothing was wool. About 10,000 to 15,000 years ago, people began to tame wild animals. These animals, such as sheep and goats, became domesticated over time. (A *domesticated* animal is an animal that, through generations, has become used to living with people; its young will be domesticated, too.) People soon realized that they could use the wool from sheep or goats to make *felt* for clothing. In this process, the wool was cut from the animal and wetted. The wet wool was then *pummeled,* or hit repeatedly, causing the wool strands to stick together, or mat. Pieces of the resulting felt were then sewn together to make clothing and blankets.

Another change in clothing occurred when people learned that they could plant seeds that would yield useful crops. Some seeds yielded crops from which clothing could be made. People in warm climates, like that of ancient Egypt, began to weave clothing from a plant called *flax.* The cloth that they wove was called *linen,* which was much lighter, in both weight and texture, than wool. This lighter material was perfect for the warm climate of northern Africa. *Cotton* was another plant from which people made cloth.

In the colder climates of northern Europe, however, people still used animals for clothing. They discovered that wool could be twisted into threads, and the threads could then be woven into clothing. Often, people combined clothing from animals with clothing from plants, rather than using just one or the other. For example, shirts made of linen might be lined with fur, while pants were made from wool. Fur-lined gloves and shoes made of *leather,* or treated animal skin, kept hands and feet warm.

## Primitive Sewing Tools

Today, we sew items with a thin metal needle and a length of thread, both of which we buy in a store. So how did early humans sew their animal skins together? Also with a needle and thread! The needle was made out of the bone or antler of an animal, and the thread was a strand of animal tendon. (A *tendon* is a body tissue that connects a muscle to a bone.)

## Not Fit for a Mummy

Did you know that the ancient Egyptians preferred to wear clothes made from plants like linen and cotton rather than those made of animal skins or even wool? The Egyptians thought that clothing made from plants was cleaner than clothing made from animals. They also thought that linen and cotton cloth were signs of civilization. In their eyes, people who wore animal skins were not civilized enough to plant farms and form settlements. Ancient Egyptians placed such a high value on clothing from plants that mummies were wrapped in linen and cotton cloth, too.

 ## MORE THAN JUST PROTECTION

As civilizations and populations grew, clothing was used for different purposes. It was still worn to protect the body, but it was also worn to show one's status in society. For example, people during ancient Roman times wore long flowing swaths of cloth called *togas*. If the edge of the toga was dyed purple, it meant that the person was important in Roman society.

Throughout history, clothing continued to be a symbol of one's status or place in society. Even if fur was not needed for warmth, it was worn as a fashionable accessory. For example, wealthy people in Europe during the Middle Ages (from the sixth to the fifteenth centuries) wore tunics trimmed in fur. The purpose of the fur was not so much to keep the person warm, but to dress up the tunic and make it "fashionable."

Wearing the latest fashions proved that one was able to afford new clothing. This is still true today. People often want the latest pair of sneakers or a shirt with a particular logo, even though these items may be more expensive than other items of the same or similar quality.

#  ANIMALS AND CLOTHING TODAY

Animals are still used for clothing today. Here are some ways that animal products are used in clothing or as accessories.

- Wool from sheep and goats is still used to make sweaters, blankets, and other clothing items.

- Leather from animal hides is still used to make jackets, coats, shoes, pants, and many other items.

- Fur from a variety of animals is still used to make coats and hats.

- Feathers from ducks and geese are still used to fill jackets and blankets.

- Ivory from elephant tusks is still being made into jewelry.

- Silk from cocoons of silkworm moths is still used to make clothing.

Many people question the use of animals for clothing today. They point out that since many types of clothing can be made from either plants or human-made products, there is no longer a need to hunt or even to raise animals for clothing. They feel that the rights of the animals should be protected, that no animals should be used for clothing at all, no matter how well the animals may be treated. People who wish to protect animals from being used for any purpose are called *animal rights activists* or *advocates*. (An *advocate* is someone who supports or defends a cause.) Most animal rights activists are also against using animals for food, entertainment, and medical research.

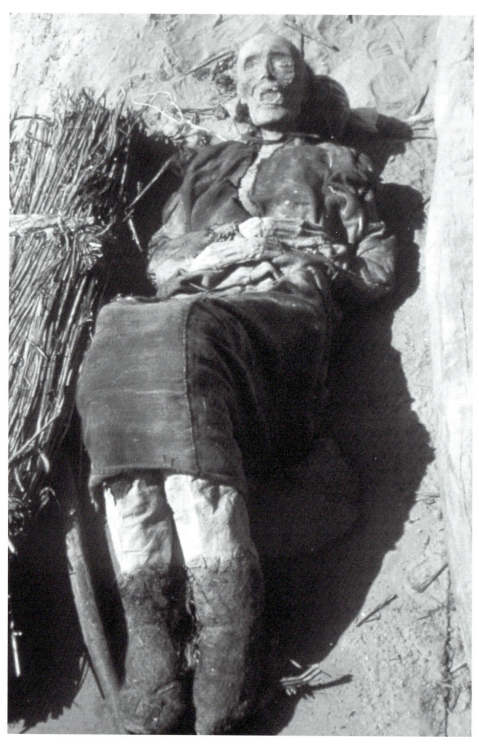

*This mummified woman, who was excavated in a Chinese village, is still wearing the colorful robes, boots, stockings, and hat that she took to her grave 3,000 years ago.*

However, other people argue that it is okay to hunt, trap, or raise animals for clothing, as long as the animals are treated humanely. These people are called *animal welfare advocates*. They point out that animals have been used for clothing for thousands of years. They say that it is a human right to use animals, as long as people don't *misuse* animals. Like animal rights activists, animal welfare groups are concerned about proper animal treatment. However, unlike animal rights activists, animal welfare groups feel that people can and should use animals for human purposes, including for clothing, food, entertainment, and research.

Today, the fashion industry is big business. The clothing that animals provide is a large part of that business. Many people in the clothing industry argue that using animals to make clothing materials provides jobs for farmers, scientists, and many others. Using animals for clothing, therefore, benefits not only the people who buy clothing made from animal products, but those who work with the animals.

As you read the following chapters in this book, you'll see both sides of the coin where animal rights are concerned. You'll be given the views of both the people who raise animals for clothing and the people who want to protect the animals from being used in the clothing industry.

## CHAPTER 1

# The Fur Trade

It may be hard to imagine, but 50,000 years ago, when early humans inhabited Africa, no humans lived on the continents of North and South America. Early humans began to travel northward from Africa and eastward across the Asian continent, usually in search of food. Eventually, their search led them to a new continent, which they entered by crossing a strip of land between the Arctic Ocean and the Bering Sea.

This is how scientists believe that people came to inhabit the North and South American continents. These early people were the ancestors of today's Native Americans.

Over time, the planet started to warm up, so glaciers began to melt and the oceans became bigger. The land bridge over which these people had crossed from Asia was eventually covered by water. The people who now lived in North and South America were essentially cut off from the rest of the world. It would be thousands of years before anyone else discovered not only the people who lived here, but the animals and other resources of these lands.

##  ENTER THE EUROPEANS

Christopher Columbus was just one of many Europeans interested in exploring the lands of North and South America. The motivation for many explorers was the search for treasure. Some were looking for cities of gold. Others, however, were satisfied with *commodities* (products) that the people native to the land seemed eager to trade—one of which was animal fur.

Christopher Columbus arrived in America in 1492. Forty-two years later, in 1534, a French explorer named Jacques Cartier came to North America, to land that today is part of Canada. Cartier's first encounter with the native people was with those of the Micmac nation. The Micmac presented animal furs to Cartier, and the fur trade had begun. People from England, Spain, and France were quick to trade things that the native people could not easily make themselves, such as items made from metals like copper and iron. Nearly seventy years before the first permanent settlement in the Americas (Jamestown, Virginia, in 1607), Europeans understood the wealth of resources that this "new" land had to offer.

## Beavers

One of the first animals involved in the fur trade was the beaver. Initially, there was little desire for beaver fur. The European fashions of the early and mid-1500s called for furs imported from Russia. In the late 1500s, however, fashions changed, and the latest craze called for hats made of felt. Beaver fur was perfect for felt making, for it could be easily matted. Not only were beavers plentiful on the North American continent, but the native peoples were more than happy to trade the beaver skins.

In the 1500s, beavers could be found in most rivers, streams, and ponds. A beaver's body is perfect for swimming. Its fur is long

### More Beaver Bits

Did you know that the beaver is actually part of the rodent family? In fact, it is one of the largest rodents in the world. The beaver has webbed feet and a flat tail, which make the animal a natural swimmer. When a beaver dives underwater, it closes its ears and nose to keep water out. Beavers also have a layer of fat under their fur, which keeps them warm in cold waters.

and soft, and its brown color helps the beaver to blend in with its habitat. The beaver also has two oil glands. It rubs the oil from one gland through its fur, which makes the fur waterproof. The oil from the other gland, called *castoreum,* has a scent that is used by the beaver to mark its territory. Along with beaver fur, people also sought castoreum, which they used in perfumes.

Because the beaver population in North America was so large, many Europeans went there to trade for and hunt the animal for the European fur market. In fact, at one time, there were hundreds of thousands of beavers in North America. As the centuries passed, however, beaver populations became scarce, and the animal almost became extinct. (An *extinct* animal species is one whose population has completely died out.) The beaver has slowly made a comeback, but its numbers are nowhere near what they once were. Today, the beaver population is estimated to be about 2.5 percent of what it once was.

## ☺ THE FUR TRADE AND FURTHER EXPLORATION

In order for beaver skins to get to Europe, they needed to be transported there by boat. During the 1600s and into the 1700s, one of the main ports for exporting the skins was Montreal. Montreal was founded in 1642, and until 1760, it was a city in the French colony of New France. (Today, Montreal is part of Canada.) Here, native peoples would often bring their animal furs, trading them for such goods as copper pots, iron ax heads, and glass beads. For the Europeans, the situation seemed ideal and profitable.

However, some Europeans were impatient and greedy. They wished to hunt and trap the beaver themselves, instead of having the natives do it. After all, why trade for the beaver furs if they could hunt beavers themselves? French fur traders soon embarked in canoes to find their own beaver populations. Called *coureurs de bois,* or "woods rangers," the French trappers traveled deep into the North American continent in search of beavers to bring back to Montreal. By the late 1600s, they had explored as far west as the Great Lakes and the Mississippi River.

## Sea Otters

Exploring North America eventually brought Europeans as far west as they could go—the West Coast. In the Pacific Northwest, another animal was discovered whose fur was sought by the Chinese—the sea otter.

Like the beaver, the sea otter has fur that is soft and thick. Considered a marine mammal, because it spends most of its life in salt water, the sea otter lives within 1 mile (1.6 kilometers) of the coast. It stays near the land for several reasons. The land serves as shelter when the weather turns rough. Also, sea otters need food normally found along the ocean's bottom, such as

*A sea otter munches on shellfish as it swims in Valdez Harbor in Alaska.*

crabs and shellfish. The otters must dive down to reach the food, and areas closer to shore are usually shallower than the open seas. Sea otters can only dive to depths of about 20 to 30 feet (6 to 9 meters).

Unfortunately, sea otters were hunted nearly to extinction. Living so close to shore made the otters easy for hunters to catch. Also, unlike most marine mammals, which spend a large portion of their time underwater, sea otters spend most of their time floating on their backs on the surface of the water. Because they were so visible, sea otters were especially easy to find.

Massive hunting for sea otters began in the 1780s. It took only thirty years before they were nearly extinct. Because so few sea otters were left, the selling and trading of sea otter fur came to an end in about 1815. Today, sea otter populations are protected—it is illegal to hunt or kill sea otters in most places around the world.

## Bison

One of the animals most often associated with North America and Native Americans was the buffalo. The proper name for buffalo is actually *bison*. Scientists estimate that before European settlers arrived in North America, the bison numbered in the millions. They lived mainly on the grassy plains in the western part of the continent, grazing on grass and resting to chew their cud, as cows do.

Bison were particularly important to the native peoples of the plains. They used bison meat for food, bison fur for clothing, and bison hide for tepees. Bison were so plentiful, and the Native Americans were so respectful of them, that the bison population was not in danger at this time. The Native Americans killed only what they needed to survive.

As European settlers began to travel westward across North America, they, too, began to hunt bison. Sometimes they traded with the Native Americans for bison skins, but they also killed thousands of bison themselves. Although European settlers mostly killed bison for their hides, they sometimes killed the animals for sport. The Europeans introduced several things to Native Americans that made hunting the bison easier—horses and guns. It was much easier to chase the bison on horseback than on foot, and it was much easier to kill the bison with guns than with arrows or spears.

The height of the bison-hunting period was the mid-1800s. By 1889, less than a thousand bison were left. In 1905, the American Bison Society was established in conjunction with the

*This is a 1966 photo of the descendants of the few hundred buffalo that survived the "Great Kill" that ended in 1889. Buffalo are an important part of the balance of life for many Native American tribes.*

## More Bison Bits

Did you know that bison actually belong to the same animal family as goats and sheep? These animals are *bovids*. Bovids have horns and hooves with similar characteristics. They eat mostly grass. Bovids also have a special stomach that allows them to swallow grass and then bring it up again to chew as cud.

Bronx Zoo. Its purpose was to replenish the bison population. It did so not only by protecting the animal from further hunting, but by setting aside lands for the bison to graze on. Today, the bison population is estimated to be 200,000.

##  ON TO THE PRESENT

Exploring the way that animals were used for clothing in the past can teach some lessons about what not to do in the future. Although beavers, sea otters, and bison are still here today, they were once on the verge of extinction. The intense hunting of the past is no longer common in today's world, but many animals are still killed for their fur. The modern ways of using animals for clothing are still a major concern for many animal rights and animal welfare groups.

## CHAPTER 2

# Fur Farms: Mink

As it was centuries ago, the fur industry today is big business. According to the Fur Information Council of America (FICA), sales of furs in the United States in 2000 were over $1 billion. Although the increase in fur sales during the 1990s was minimal, the jump in fur sales from 1999 to 2000 was an increase of more than 20 percent ($1.4 billion to $1.69 billion).

Although these numbers may be music to the ears of the fur industry and fur-industry supporters, they sing a different song for animal rights groups. For many animal rights advocates, the death of the fur industry could only mean life for the animals that they try to protect.

## WHAT IS A FUR FARM?

Typically, animals raised for food are raised on farms. Farms are specialized places where the animals are bred, born, raised, cared for, and then eventually killed and turned into food products.

Animals raised for fur are also raised on farms, which are also called ranches. Just as on food farms, the animals on fur farms are bred, born, raised, cared for, and then eventually killed, only not for their meat, but for their fur. The animal most often raised on a fur farm is the mink.

According to the Fur Commission USA (FCUSA), a nonprofit organization that represents mink and fox farms in the United States, fur farming is largely a family-run business. In fact, not only are most fur farms in North America family businesses, but they have belonged to the families for several generations.

Fur-farm associations claim that their animals are treated humanely. After all, they explain, a healthy animal produces healthy-looking fur. However, animal rights activists and animal welfare groups are both quick to point out many ways in which animals that live on fur farms are not only mistreated but denied any quality of life.

To clarify why conditions on fur farms might not be in the best interest of the fur-bearing animals, here is a quick look at how mink live in the wild.

##  MINK IN THE WILD

Mink are part of the same animal family to which weasels, otters, and skunks belong—*mustelids.* Mink have long bodies and tails and short legs. They can grow to 25 inches (63 centimeters) long, not including the tail. Not only are they good swimmers, like their otter cousins, but they also climb trees. Mink are usually solitary animals, preferring to live on their own rather than with a group of other mink.

Mink are meat-eaters, or *carnivores,* and they prey on a variety of animals. When hunting in the water, mink eat frogs, crayfish, and minnows. When hunting on land, mink kill rabbits, birds, snakes, and other small creatures. Likewise, mink in the wild are hunted by other animals, such as coyotes, bobcats, and owls, for food.

Mink usually mate in late winter, and the females give birth in the spring, usually to between two and ten babies. The female mink cares for her young, and at the end of the summer, the young mink are usually able to fend for themselves. Mink are considered adults when they are about a year old, at which time they are able to reproduce. In the wild, mink usually live until they are about three years old, although they may be killed earlier by predators.

# MINK ON THE FARM

Life on the farm for mink is very different from the life that they would normally live in the wild. Instead of streams to swim in and trees to climb, mink are usually sheltered in small cages. Despite the fact that mink are by nature solitary animals, they are usually housed with one to three other mink.

Fur farmers explain that their animals receive the highest quality care. However, animal activists point out that raising the mink in cages deprives the animals of their natural tendencies. In addition, according to animal activists, mink sometimes resort to biting themselves. This behavior, they claim, is in reaction to their confinement.

*This mink stands on its hind legs and peers from its cage.*

*In central England, a mink farm worker releases minks back into an enclosed compound on a mink breeding farm in September 1998. Almost 8,000 minks were freed here by animal activists who hacked their cages open.*

Mink raised on farms are usually killed when they are about a year and a half old, although some mink live a few more years if kept for breeding purposes. The FCUSA describes the procedure for killing mink as "humane euthanasia," and points out that its technique is recognized by the American Veterinary Medical Association (AVMA). (*Euthanasia* is the act of painlessly killing an animal.) As the FCUSA explains on its Web site, "The animals are placed in a special airtight container which has been prefilled with gas. The unit is mobile and is brought to the cages to minimize stress from handling. The animals are immediately rendered unconscious and die without stress or pain." The gas that the farmers use is carbon monoxide or carbon dioxide bottled gas.

Animal activists describe the process in a different way. One such animal rights group is the People for the Ethical Treatment of Animals (PETA). On its Web site, PETA claims, "Small animals can be shoved up to 20 at a time into boxes, where they are poisoned with hot, unfiltered engine exhaust pumped in by hose from the fur farmer's truck." They also charge that the gas is not always effective, and sometimes the animals are only sleeping and not dead. These unfortunate animals, PETA asserts, wake up as they are being skinned.

Despite these claims by PETA, the FCUSA insists that its animals are treated with the utmost care. To encourage animal welfare practices on fur farms, the FCUSA has established the Merit Award certification program. Fur farms must apply for the certification seal, at which time they are inspected to ensure that all animals are treated humanely. As the FCUSA notes on its Web site:

The Merit Award seal is an honor for commitment to humane treatment in all aspects of fur farming:

- Vigilant attention to nutritional needs

- Clean, safe, and appropriate housing

- Prompt veterinary care

- Consideration for the animal's disposition and reproductive needs

- Elimination of outside stress

For a fur farmer to receive the award, the farmer must feed the animals healthful food, keep them in clean, safe places, and alert a doctor when the animals get sick. The farmer must also pay attention to the animals' personal needs, as well as what the animals need in order to produce offspring. In addition, the farmer must also make the animals' lives as stress-free as possible.

## Clarifying Animal Rights vs. Animal Welfare

People for the Ethical Treatment of Animals (PETA) is an animal rights group. It believes that it is an animal's right not to be used by people for any purpose, including for clothing. PETA often points out the mistreatment of animals when animals are being used for a specific purpose, such as for clothing. PETA's goal, however, is not to improve the treatment of animals, but to stop the use of animals altogether. On another side of the debate are animal welfare advocates. People who work with animals, including fur farmers, support animal welfare. Animal welfare differs from animal rights. Animal welfare means that animals can be used by people, as long as the animals are treated well—as long as their *welfare* is a priority.

##  MINK PRODUCTS

The main product used from the mink is its fur. In fact, people often think "mink" when they hear the words "fur coat." Mink are probably the most popular type of animal associated with fur products. But mink are not very big animals—most are 20 to 25 inches (50 to 63 centimeters) long. Therefore, it takes not one, not two, not even three, but several dozen mink to make one fur coat! Although fur industry supporters point out the warmth and luxurious texture of the fur coats, animal rights activists wonder how anyone could feel comfortable wearing dozens of dead animals.

Fur isn't the only product that comes from mink. According to the FCUSA, a mink's body can be processed into "protein meal," which is a basic ingredient in pet and animal food. Mink have a layer of fat between their skin and their body from which comes oil. The oil is used in such products as

*A model shows off a new mink coat at a fashion
show in New York City in September 1966.*

hypoallergenic soaps, cosmetics, and hair products. Manure produced by the mink can be used as fertilizer for crops.

##  MINK BY THE NUMBERS

The number of farm-raised mink is monitored by the United States Department of Agriculture (USDA). Wisconsin was the top breeder of mink in 2000, followed by Utah. Other states that have mink farms are Idaho, Illinois, Iowa, Michigan, Minnesota, New York, Ohio, Oregon, Pennsylvania, South Dakota, and Washington.

On a global scale, the United States raises about 10 percent of the mink in the world, and Canada raises another 4 percent. The countries that raise the most mink for fur production are those of Scandinavia, which combine to create 56 percent of the world mink market, according to the FCUSA.

During the 1990s, mink pelt production was actually on the rise. (A *pelt* is an animal skin with the fur still attached.) In

---

## Mink-Producing States

The numbers below indicate the number of mink *pelts* (skins with the fur still attached) produced in the United States in 2000, as reported by the United States Department of Agriculture (USDA).

| State | Number of Mink Pelts |
|---|---|
| Wisconsin | 680,100 |
| Utah | 590,000 |
| Minnesota | 284,800 |
| Oregon | 268,000 |
| Idaho | 222,400 |
| Other states | 620,800 |

---

1993, the total number of mink pelts produced was about 2.5 million. By 1998, that number had risen to almost 3 million. Then mink pelt production began to decrease slightly. For example, in 1999, according to the USDA, the total number of mink pelts produced in the United States was 2,812,500. In 2000, that number had gone down to 2,666,100. According to the USDA, the number of mink farms is also on the decline. In 1987, there were over 1,000 mink farms in the United States. By 1999, that number had dropped to 398.

These numbers, although discouraging to the fur industry, are encouraging to animal rights advocates. Fewer fur farms mean fewer animals being born and raised for the purpose of being turned into coats and hats. Animal rights advocates hope that these declining numbers demonstrate that society has begun to have a conscience about wearing animal fur.

Even though the number of mink farms has gone down, other animals, such as rabbits, foxes, and chinchillas, are also used for fur. In addition, fur is imported from overseas. These facts could contribute to the rise in fur sales in the United States, as described at the beginning of the chapter.

CHAPTER 3

# Fur Farms: Other Animals

One reason that farming fur is more desirable to the fur industry than hunting the animals in the wild is that farmers can affect the quality of the fur that they produce by *controlled breeding*. In controlled breeding, the farmers choose which animals to breed with each other to produce offspring with certain traits. By choosing animals that have fur with desirable qualities, such as a particular color or texture, farmers are able to ensure that the young will have equally desirable fur. In addition, farmers may also *crossbreed* animals—mate animals of the same species but of different varieties—to produce new or unusual fur. The first such fur developed through crossbreeding was silver fox fur, which was derived from the red fox.

Although mink are the most common animals raised for fur, they are not the only one. Foxes are the next most common animal found on fur farms. Other fur farm animals include chinchillas, sables, and rabbits.

##  RED FOXES IN THE WILD

Red foxes can be found not only in most parts of the United States, but also in countries around the world, including those of Europe and Asia. Part of the dog family, red foxes are *omnivores,* which means that they eat both animals and plants. Along with eating such animals as rodents, rabbits, and even insects, foxes will also munch on berries and other plant life. Red foxes are skilled hunters, listening for their prey and then

*This wild red fox warily eyes the camera while out for a morning walk.*

pouncing on it when the prey gets near.

Red foxes usually mate in the winter, and the young are born some fifty to sixty days later, usually in mid-March. From four to eight pups may be born at a time. The pups are usually born in an underground den, and the mother often stays with the pups while the father hunts and brings back food.

During August and September, the young foxes begin to leave their family units to search for mates and territories of their own. Red foxes are fast animals that can run up to 45 miles (72 kilometers) an hour if they are being chased. Red foxes are curious and playful, but also shy and nervous. In the wild, a red fox may live for twelve years.

#  RED FOXES ON THE FARM

Like farm-raised mink, farm-raised foxes live very differently from foxes that live in the wild. One to four foxes are kept in a cage that measures about 2.5 feet (0.75 meters) square. Confined to such cages, the foxes are not permitted to follow their natural instincts to hunt and explore. Their lives are dull and meaningless. Since they are such intelligent animals, animal advocates point out that such confinement, devoid of any activity, is emotionally stressful to the foxes. Investigators and handlers admit that farmed foxes are often anxious and fearful. An even more disturbing result of the farm experience is that some female foxes resort to *infanticide*—the killing of their own babies.

Of even more concern to animal rights and welfare groups is the means by which foxes are sometimes killed. Photographs on animal rights Web sites show foxes with their mouths held closed by painful clamps. A rod is then inserted into the animal's rear end, or *anus,* and the fox is then electrocuted until dead. Sometimes, the fox may also be poisoned. The farmer's goal is to kill the animal in such a way that its fur is not ruined.

## Fox Farms Revisited

In light of fox farm investigations, many countries around the world are revisiting their views of fox farms. Sweden is trying to pass laws that ensure that foxes kept on fur farms will be housed in a more humane manner, with room to move, dig, and be active. Austria has banned fox farming altogether. The Netherlands also plans to phase out fox farming. In Denmark, the Danish Ministry of Agriculture's Ethical Council for Domestic Animals stated that fox farming is ethically unacceptable.

The FCUSA claims that the only approved method of killing foxes is by lethal injection, which causes cardiac arrest. According to its Web site: "Lethal injection is the only fox harvesting method recommended by the FCUSA Animal Welfare Committee." However, the Humane Society of the United States (HSUS), an animal welfare group, points out that no laws exist to govern how animals are handled or killed for fur. The HSUS also explains that the recommendations put forth by the fur industry on how to care for the animals are *voluntary*—the fur farmers don't have to follow the recommendations if they don't want to do so.

##  MORE FUR-BEARING ANIMALS

Along with the mink and the fox, several other animals are popular on fur farms.

### Chinchillas

Like many animals hunted for their fur, chinchillas are nearly extinct in the wild. Small animals, they measure 8 to 15 inches (20 to 38 centimeters) long, with a bushy tail 3 to 6 inches (8 to 15 centimeters) long. In the wild, chinchillas live in South America, mostly in the Andes Mountains. They often live in large colonies of 100 or more. Chinchillas are not only raised on farms for their fur, but recently have been sold to people as pets.

### Sables

Sables are similar to mink in that they have long bodies and tails and short legs, although overall, they are slightly smaller than mink. Also like mink, sables have beautiful fur, which is highly prized in the fur market. In the wild, sables typically live in the forests of Siberia. However, overhunting of sables for their fur has greatly reduced their numbers, and conservation efforts have been put in place to protect and restore wild sables. Sables are also bred, born, and raised on fur farms.

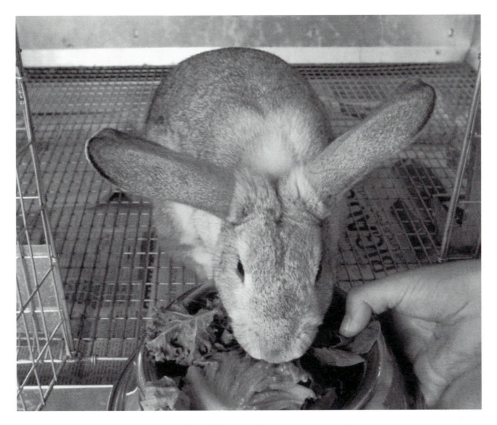

*This gray chinchilla rabbit gets its name from the similarity of its fur to that of a chinchilla.*

## Rabbits

Rabbit populations in the wild never seem to be on the decline. However, to control rabbit-fur production, many rabbits are raised on fur farms. Rabbits, of course, are also kept as pets.

##  A SHOCKING DISCOVERY

In the late 1990s, perhaps one of the most shocking discoveries for many people in the United States was the use of dog and cat fur in fur products. People in the United States and in many other countries around the world love their pets, specifically

*In March 2000, Florida state senator James Sebesta holds up toys made from dog and cat fur as he leaves a panel considering his bill to outlaw the use of pet fur for such products.*

cats and dogs. In fact, in the United States alone, sources estimate that 68 million dogs and 73 million cats are owned as pets. People spend a lot of money caring for their pets, and they often treat their pets as members of their own families.

Maybe that is why in 1998, Americans and other people around the world were horrified when they learned that loving cats and dogs were being raised and killed in Asia for their fur. Perhaps most shocking of all, however, were the ways in which the dogs and cats were not only treated, but killed.

In the 1990s, the HSUS, along with Humane Society International (HSI) and a German journalist, went undercover in China to investigate claims that dogs and cats were being killed for the fur industry. The claims proved to be true. The investigators estimated that about 2 million dogs and cats were being killed each year for their fur.

## A Fate Most Cruel

Investigators found that dogs and cats are raised on what could be identified as fur farms. A dog farm might have 300 dogs, while a cat farm might have 70 cats. Dogs on these so-called farms were kept in cold, dark buildings without food or water. Most were tethered by thin metal wires, and some were kept in sacks. The process by which the dogs were killed was particularly cruel, with the dogs being stabbed and left to bleed to death. Most of the dogs killed were German shepherds. It takes about ten to twelve dogs to make one coat.

The lives of cats on cat farms are no better. Living in small, bare cages, the cats are eventually killed by hanging or other cruel means. The most sought-after cat fur is that of short-haired, gray cats and short-haired, orange tabby cats. It takes about twenty-four cats to make one coat.

## No Toying Around

Cat and dog fur has shown up not only on clothing, but on toys and figurines, too. These toys are usually small figures of cats or dogs—they have fur so soft that it feels like the real thing. Well, in some cases it is. The Humane Society of the United States (HSUS) revealed that dog and cat fur was being used for such figurines. The labels on the figurines don't say what type of fur is used, but usually declare that the products were "made in China."

In 2000, nearly two years after the HSUS revealed the findings of its investigation, President Bill Clinton signed a ban on dog and cat fur. It is now against the law in the United States to import, export, or sell items that use dog or cat fur.

## I'd Never Buy That!

Many people who buy fur products claim that they would draw the line at products made with dog or cat fur. However, sometimes the fur is not labeled properly. According to the HSUS, dog fur may be labeled as "gae-wolf," "sobaki," or "Asian jackal." Cat fur may be labeled as "wildcat," "goyangi," or "katzenfelle."

Although full-length coats of dog and cat fur have not made their way to the United States, smaller fur items have. For example, dog or cat fur has been found in the fur trim on hats, gloves, and coats. The only way to prove from which animal the fur comes is to conduct DNA tests. When DNA testing was done on several items trimmed with fur that was supposed to come from a more "exotic" animal, the tests proved that it was indeed dog fur.

#  A STRANGE CONTRADICTION

Many Americans were outraged when they learned about the cruelties inflicted on dogs and cats for the purposes of the fur market. However, fewer people seem concerned over the plight of other animals raised on fur farms here in the United States. It seems like a strange contradiction—the lives and welfare of cats and dogs appear to be more important than the lives and welfare of other animals.

Animal rights activists make no such distinction. They view the harm inflicted on all animals as unnecessary and cruel. Pet owners love to see their animals happy and living as freely as possible. Animal rights groups claim that all animals should have the same freedoms. Being bred, born, and raised on fur farms does not allow mink, foxes, and other fur-bearing animals a chance to live freely and naturally.

## CHAPTER 4

# Fur Trapping

Not every animal used for fur is raised on a fur farm. Many wild animals are also killed for their fur. Although the practice of capturing animals for their fur dates back thousands of years, many people today disagree with this practice. Even so, over half the animals used for the fur market are wild animals that have been caught and killed by fur trappers. The methods by which these animals are caught lead many people to question the methods used in fur trapping and to raise the issue of animal rights and animal welfare.

## WHICH WILD ANIMALS ARE TRAPPED FOR FUR?

It is estimated that between 3.5 million and 4 million wild animals are trapped and killed for their fur each year in the United States. Of that number, the most common animal killed, according to the HSUS, is the raccoon. On average, over 2.5 million raccoons were killed per year in the United States from 1995 to 1998. That's nearly as many as mink killed on fur farms, which was 2.6 million per year during the same time period.

The raccoon is a highly recognizable animal, found throughout much of North and Central America. In fact, raccoons have learned to adapt very well to human populations, which encroach on forests and other natural raccoon habitats. Normally active at night, raccoons eat a variety of foods, from small animals and insects to such plant foods as berries and nuts. With their versatile diets, raccoons can sometimes be found scavenging through suburban garbage cans.

Raccoons have several distinguishing physical features. A black swath of fur across the eyes gives the animal a "masked" appearance. Its long, fat tail is striped with rings of black and light gray. The digits on each paw are extremely adept at handling food and climbing trees.

Raccoons are not the only animals trapped for their fur. The table below lists fifteen animals identified by the HSUS, including raccoons, that were killed for their fur from 1995 through 1998. Although mink and fox are raised on fur farms, a large number of these animals are also caught in the wild, so they are listed in the table as well.

| Animal | Number Killed in the Wild |
| --- | --- |
| beaver | 291,000 |
| bobcat | 40,000 |
| coyote | 491,000 |
| fisher | 6,000 |
| fox | 368,000 |
| lynx | 1,200 |
| marten (American sable) | 12,000 |
| mink | 115,000 |
| muskrat | 1,358,000 |
| nutria | 237,000 |
| opossum | 211,000 |
| otter | 20,000 |
| raccoon | 2,575,000 |
| skunk | 63,000 |
| weasel | 7,000 |

*A pair of young raccoons peer through roadside grass in Buzzle Township, Minnesota.*

## HOW ARE THEY CAUGHT?

Traps to catch fur-bearing animals come in two categories: those that kill the animals and those that hold the animals. Fur management groups and animal activists differ about the consequences of each.

The most common type of trap is the *leghold trap*. When an animal unwittingly steps into this trap, a spring is released, and the trap snaps around the animal's leg. The animal is trapped, unable to release itself from the leghold, while at the same time, the animal's fur is kept intact.

*A red fox lies caught in a leghold trap in Charlotte, Vermont. Trappers are participating in a study of new types of traps that are designed to cause less injury to animals.*

The most popular type of "kill" trap is the *body-gripping trap,* or Conibear. This trap is designed not to hold the animal by a leg, but to kill it almost instantly. When the animal steps into this trap, metal bars snap down around the animal's body, breaking its back or neck.

A *snare* is another type of trap. This trap uses a cable shaped like a noose, which traps an animal around the neck or a leg. As the animal struggles, the snare tightens. The animal either dies from lack of oxygen or merely is caught to wait for the trapper to arrive.

## WHAT ANIMAL ACTIVISTS SAY

Animal activists claim that all three of these traps cause immense pain and injury to the animals. Leghold traps can tear and break the animal's skin and bones; the animal can break its teeth trying to bite through the trap or actually gnaw off its own foot in order to escape. Animals caught in kill traps often do not die right away, but remain alive as they suffer broken backs or necks. Snare traps are also painful. A snare that catches around the neck will eventually choke and strangle the animal; a snare that catches around a leg will eventually cut off blood flow, causing damage to the limb.

According to animal activists, this suffering may go on for one to three days, because the animals must remain in the traps until the trappers arrive. During that time, an animal may die of starvation, dehydration, or freezing. Also, because the animals cannot run to escape predators, many captured animals become easy prey themselves.

Once trappers do arrive to check on their traps, the animals are killed. Usually, the animals are clubbed on the head until dead. To animal rights and welfare groups, the traps and the manner in which these wild animals spend the last hours or days of their lives are cruel and inhumane.

## WHAT FUR TRAPPERS SAY

Fur trappers refute these claims. They say that regulations require trappers to check their traps once a day, so the animals do not languish for days there. They also say that traps are constantly being improved to make the trapping experience as painless and stress-free as possible. For example, some leghold traps now have padded jaws to make the initial impact softer, thereby reducing injury. Further, according to the National Trappers Association (NTA), the practice of clubbing an animal to death is a method recognized by the AVMA as "humane euthanasia."

The AVMA does recognize clubbing, when performed correctly, as a quick method of death for small laboratory animals. However, according to the American Society for the Prevention of Cruelty to Animals (ASPCA), the NTA might be distorting the idea of this approved means of humane

## Fifties Fashion Craze

In the 1950s, a popular fashion craze was wearing raccoon fur—on your head! "Coonskin hats" were usually brownish-gray, with the raccoon's tail hanging from one side. Coonskin hats are most often associated with Davy Crockett, a frontiersman and politician who died at the battle of the Alamo in 1836. Crockett's hat was made of real raccoon fur, as were many hats made during the 1950s. However, many of the 1950s hats were probably made with artificial materials.

euthanasia to suit its purposes; animals trapped for fur are not the same as small laboratory animals.

The trappers' ultimate goal is to obtain the animals' skins, unharmed and in good condition. Therefore, it is only to their benefit to make capturing and killing the animals as humane as possible. It is counterproductive for the trappers not to attend to trapped animals immediately and with care.

Even so, some states have made it illegal for trappers to use leghold traps. The traps are banned in New Jersey, Rhode Island, Florida, Arizona, Colorado, and Massachusetts. Eighty-nine countries around the world have also banned the traps. Kill traps have been outlawed in a number of states as well, as have snares.

##  OTHER TRAP CONCERNS

Animal activists have other criticisms of the fur-trapping industry. They claim that many "nontarget" animals are also caught in the traps, such as deer, birds, and even cats and dogs. Although these animals may be released in time, many, the activists claim, do not survive the injuries that they get from the traps. Some of these hapless victims belong to endangered species, like eagles and wolverines. (An *endangered* species is one that is in danger of completely dying out, because existing populations are very small.)

Again, the trappers paint a different picture. A rebuttal, or response, on the NTA Web site notes that only 3 percent of all animals captured were "nontarget." The NTA also points out that none of these nontarget animals were threatened or endangered. (An animal species that is *threatened* is one that may soon become endangered if its numbers continue to decrease.) The group also points out that since the leghold traps are technically "live-capture restraining devices," any animals caught can be released. In fact, according to the NTA, otters caught mostly in leghold traps were released and reintroduced into the wild in eighteen states. When an animal is *reintroduced,* it is released in an area where populations of that animal once lived, but no longer do.

## TRAPPING AND THE ENVIRONMENT

Fur trappers say not only that fur trapping is a traditional way of life and a practice that has been followed for thousands of years, but that fur trapping is also good for the environment. If managed properly, fur trapping controls animal populations that have become too large because of a lack of natural predators, such as bears and mountain lions. Fur trapping, therefore, is actually an adequate means of conservation and preservation.

One way that fur trappers ensure conservation and animal welfare is by promoting Best Management Practices, or BMPs, a set of rules for trapping "that make a trap and trapper function safely, humanely and efficiently." Working alongside state fish and wildlife agencies, trappers are trying to develop and promote the most efficient ways of capturing fur-bearing animals that follow BMPs. To meet this end, they have outlined five key issues related to animal traps.

- animal welfare
- the efficiency of the traps
- the selectivity of the traps
- the safety of the traps for people
- the practical application of various traps

According to the Northeast Furbearer Resource Technical Committee,

> Everyone—managers, regulators, biologists, veterinarians and the public who traps—is interested in using the best technology available for the responsible capture of furbearers. Working towards this goal state wildlife agencies will continue their efforts in trap research and developing Best Management Practices. Making sure trapping is conducted humanely and responsibly is extremely important to us all. Best Management Practices based on sound scientific data and biology will continue to measurably improve the welfare of captured wildlife in the United States.

This means that people involved in trapping animals for fur are always trying to improve trapping methods. In the end, these improvements will be best for the animals.

## ⊚ TO TRAP OR NOT TO TRAP?

Animal activists don't buy the argument that trapping is ecologically sound. They point out that animal populations are more adept at controlling their own numbers than fur trappers. In fact, they say that the traps often catch healthy animals that are active and looking for food. Animal populations would be much better served, therefore, if these healthy animals were allowed to live naturally.

To get their point across, animal rights groups are quick to

## What Is Selectivity?

Trappers usually try to catch specific animals, and therefore, many considerations go into the trapping process. These trapping considerations are called *selectivity*. Considerations include:

- the location of the trap

- the type of trap used

- the size of the trap

- the *trap tension,* or what will set off the trap

- the bait

- the season in which the traps are set

show pictures of animals caught in traps. Fur trappers feel that the animals suffer minimally and that it is the trapper's right to hunt animals for the things that people need. Animal rights activists—and often animal welfare groups, too—argue that people don't need coats made with animal fur, and they plead with buyers not to purchase real fur coats. Fur trappers point out that fur trapping and the fur market are important industries, providing a livelihood for thousands of people.

Even so, the HSUS reports that fur trapping in the United States is down. In the late 1980s, there were 298,000 registered fur trappers. Ten years later, that number had dropped to about 160,000. Also, although between 3.5 and 4 million animals are trapped each year in the United States, that number is down sharply from the 17 million animals that were trapped and killed each year in the mid-1980s.

## CHAPTER 5

# Sheep's Wool

Not all animals whose fur is used for clothing are killed. Some animals get their fur cut off and then grow it again. The sheep, from which we get wool, is one such animal.

People often visualize sheep as roaming in large flocks on green hillsides. Herd dogs, like border collies, often watch over sheep, along with a human counterpart. Each sheep "goes with the flow," following the movements of the sheep beside it, as well as the commands of the dog or shepherd. This willingness to obey makes sheep easy to control and raise. It is also what makes sheep perfect animals for the clothing market.

## RAISING SHEEP

Although there are over 200 different breeds of sheep, only thirty-five breeds are raised in the United States. According to the National Agricultural Statistics Service (NASS), the United States today raises some 7 million sheep a year. This number includes both full-grown sheep and lambs. The American Sheep Industry Association (ASIA) describes several different types of sheep farms:

- Small farms or homes that raise one to ten sheep, primarily for wool to make hand-knit products

- Medium-sized farms that raise 50 to 200 sheep

- Large farms that raise 1,000 to 5,000 sheep

*This farmer uses hand shears to remove the wool from a sheep.*
*Some farmers still shear, wash, and card wool by hand.*

## Desert Makers

Although sheep can be good for the ecology, too much grazing can be a problem. Sheep cut grass with their teeth, but they also pull the grass out by the roots. Some scientists believe that too many grazing sheep and goats might have led to the formation of the deserts of northern Africa.

Sheep have been raised for a variety of purposes throughout history. When sheep were first domesticated, they were primarily used for their milk and wool. As people began to settle in towns, the demand for meat increased, and a new purpose was found for sheep. On today's farms in the United States, sheep are used for meat and wool, but their grazing is useful as well.

Sheep have proved to be ecologically important animals. They like to eat many weeds and other plants that stifle the growth of evergreen trees. Relying on sheep to "weed out" these plants saves farmers from spraying harmful or toxic weed-killers over the land. According to recent statistics, 10,000 sheep helped new tree growth in a California forest that had burned in 1978. Statistics from other states and in Canada have also been promising.

Female sheep are called *ewes,* and male sheep are called *rams.* Ewes and rams usually mate in the fall, and babies, or *lambs,* are born five months later, in the spring. A ewe will usually give birth to only one or two lambs at a time. The lambs will nurse (drink milk from their mother) until they are about five months old. At that time, the lamb's stomach can handle the grasses and other plants that are part of an adult sheep's diet. Lambs are considered adult at about six months of age.

## What's the Crimp?

A sheep's fleece does not lie flat on its body, but rather is puffy and springy. That's because the sheep's wool grows in curls and waves. These curls and waves are called the *crimp*. The crimp allows garments made of wool to stretch. The crimp also makes wool warm, because it traps air in tiny pockets. The trapped air in the wool helps keep cold air from reaching the skin.

##  FROM SHEEP TO MARKET

One has only to look at a sheep to notice its thick, wooly fur, also called *fleece*. As with many animals' fur, a sheep's wool tends to grow all year and thicken in winter, when the weather is colder. The wool keeps the sheep's body warm. In spring, when the weather warms, the sheep doesn't need as much protection from the cold. Springtime, therefore, is usually shearing time. *Shearing* is the term applied to the cutting off of the sheep's wool.

Shearing is usually done with an electric razor, although sometimes the wool is cut off with special scissors called *shears*. The sheepshearer gently flips a sheep onto its back and clips the wool from its underbelly, between its legs, and along its neck and head. These areas are tricky to shear and are usually done first with a slower hand. Once the fleece is free from these body areas, the sheepshearer moves more quickly over the sheep's shoulders, back, and sides.

When done, the sheep's fleece has usually been removed in one single piece. The wool is then boxed or bagged and sent to a warehouse to be sold.

# PROCESSING THE WOOL

When the wool leaves the farm, it is usually transported to a warehouse. Here, wool merchants inspect it. They check for any dirt and other matter that might be in the wool, as well as the length of the wool fibers. Buyers then bid on the wool, and the highest bidders eventually buy the wool.

Once the wool has been purchased, it is washed, or *scoured*. Sheep's wool has an oily substance called *lanolin* in it. On a sheep, the lanolin helps make the wool waterproof. People who buy products made from wool, however, do not wish to have

*After it is sheared from the sheep, wool must be washed, carded, and died before it is made into clothing.*

## On a Smaller Scale

Not all wool is processed on large ranches and in large factories. Some sheep ranches have a small number of sheep. People who have small wool-producing farms may choose to wash and card the wool themselves. Here, the wool is carded by hand with special brushes.

this oily substance on them. During the scouring process, the lanolin is removed from the wool, along with any other dirt or plant matter that may be embedded in it.

Once clean, the wool is carded. *Carding* is combing the wool to straighten the wool fibers and to remove tangles and any remaining debris. In a factory, the wool is carded on large wire rollers.

Finally, the cleaned and carded wool is spun into yarn. In this process, the clumps of wool are pulled and made long and thin. The yarn is then used to create fabric, from which a variety of products can be made, including blankets, sweaters, hats, gloves, scarves, and even rugs.

Wool can also be dyed. Dying the wool can occur at several different stages. *Stock-dyed wool* is wool that has been dyed after it has been washed. *Yarn-dyed wool* is wool that has been dyed after it has been spun into yarn. *Piece-dyed wool* is wool that has been dyed after it has been woven into a piece of fabric.

##  SO WHAT'S THE PROBLEM?

Sheep are not killed for their wool, and according to the ASIA, sheep farmers strive to keep their animals as healthy as possible. The sheep are free to graze in pastures and are sheltered in warm barns with plenty of ventilation. During the birthing season, when lambs are born, many sheep farmers spend all day and night with their animals, offering assistance if needed. It all sounds rather idyllic. Could animal rights activists possibly object to producing wool?

They can and they do. Animal rights groups like PETA claim that shearing the wool from sheep is done before the sheep are ready to shed it. The sheep—which would normally need the wool covering for a few more weeks to remain warm—become cold. In the opinion of animal rights groups, shearing the sheep is like taking the clothes off people and leaving them naked to face the elements.

Animal rights groups also object to the ways in which the sheep are shorn. They claim that because sheepshearers are paid for the amount of wool that they shear, not the time that they

*Most of this ram's wool is sheared and lying on the floor around him.*

53

spend shearing, the shearers work quickly. In their rush to shear as much wool as possible, the shearers sometimes cut the sheep.

Other practices on sheep farms also raise concerns. Sheep are actually born with longer tails than people traditionally see in pictures. Sheep farmers cut off, or *dock*, a portion of the tails of young lambs. The farmers say that docking keeps the sheep clean, because it eliminates the problem of dirt becoming encrusted in the tail. Animal rights groups point out that the docking is done without *anesthesia*, or numbing before surgery. Therefore, it is painful for young sheep. Animal rights groups also object to the use of ear tags to identify lambs. Sheep farmers describe the process of tagging the sheep as no more painful than people getting their ears pierced.

Sheep farmers insist that their sheep receive the best care possible. As noted on the ASIA Web site: "Sheep are carefully shorn to protect the valuable fleece and the sheep that provides it. When held correctly for shearing, most sheep do not struggle and can be easily held and turned for their 'haircut.'" In response to animal rights groups' claims of "naked" animals, the ASIA says, "In some climates, sheep need shelter after being shorn. And yet, within a week, their fleeces have grown back to provide protection from all but the worst weather. The eight to twelve pounds [3.6 to 5.4 kilograms] of wool on each animal must be shorn every year to keep the animal comfortable and healthy."

## Don't Lose the Lanolin!

The lanolin oil that comes from sheep's fleece is not wasted. Since oil and water don't mix, the lanolin oil that has been washed out of the wool while it is being cleaned is separated from the water. Lanolin oil is used in a variety of products, including soaps, cosmetics, and lotions.

## Wool Down Under

The two countries that raise the most sheep are New Zealand and Australia. According to one source, New Zealand has a human population of 4 million and a sheep population of 50 million. Not all the sheep are used for wool, however. Sheep there are also used for their meat and milk. Australia has nearly 150 million sheep. In fact, Australia produces about 80 percent of the world's wool.

Animal rights groups disagree. "Many people believe that shearing sheep helps animals who might otherwise be burdened with too much wool," states the PETA Web site. "But without human interference, sheep grow just enough wool to protect themselves from temperature extremes. The fleece provides effective insulation against both cold and heat."

The question, then, is this: Is it the right of the sheep to keep their fleece? Or is it the right of people who care for the sheep to remove the fleece for human needs?

## CHAPTER 6

# More Fine Fibers

*Fiber* is a term to identify something that is used to make material or fabric for clothing. Fibers can be made from animals, plants, or artificial materials. For example, wool shorn from sheep is also known as *wool fiber*. From the wool fibers, fabrics can be woven, and clothing can be made. Other animals also produce fiber from which fabrics and clothing are produced. These animals include goats, alpacas, llamas, vicuñas, and camels.

## ⊚ GOATS IN GENERAL

Although goats may appear similar to sheep, they differ from sheep in many ways. A female goat is called a *doe* or a *nanny,* and a baby or young goat is called a *kid*. The male goat is called a *billy* or *buck*. He has a distinctive beard—or *goatee*—that sheep do not have. Goats also have shorter tails than sheep, and their horns grow outward, instead of growing in spirals like sheep horns do. The pupils of a goat's eyes have a different shape than those of most animals. If you look at your eyes in a mirror, you will see that your pupils are round. A goat's pupils are rectangular!

Goats are excellent mountain climbers, and in the wild, they are usually found in mountainous areas. Their feet are equipped with special hooves that make them agile at maneuvering across rocky terrain.

Goats are also notorious eaters, consuming just about anything. They eat the same plant foods as sheep, plus a whole

*The goat on the right stands in its stall near the goat from which it was cloned at Texas A&M University in September 2001.*

lot more, such as nuts, berries, and leaves. In fact, goats may even stand upright on their back legs to reach the leaves on trees. Most goats on farms are allowed to roam within fenced-in enclosures. The enclosures not only prevent the goats from leaving the area, but also protect them from predators.

People who breed and raise goats often do so for a specific purpose. For example, some people raise goats for their milk. Although not popular in the United States, goat milk is the most common type of milk that people drink in the rest of the world. Three breeds of goats often raised for milk are

Toggenburgs, LaManchas, and Nubians. In some parts of the world, people also raise goats for their meat. Some goats are also bred and raised for their wool. These goats are Angoras and goats with cashmere fur.

Typically, goats have two types of fur. One is the *primary fur,* or fiber, which is usually straight. This is the fur that you see. The other fur is the *secondary fur*, or fiber, which is usually curly. This is an undercoat. It usually doesn't grow as long as the primary fur, and therefore cannot be seen unless some of the primary fur is brushed aside. It is these secondary hairs, also called *down,* that most fabric-makers use for clothing.

## Angora Goats

If one were to look at an Angora goat from a distance, one might at first mistake it for a sheep. On closer inspection, however, one would notice the distinct characteristics that make this animal a goat, such as its horns, which grow outward, and its eyes, which have rectangular pupils.

The confusion lies in the fact that the Angora goat has a long coat, which from a distance looks like the wool coat of sheep. Up close, Angora wool and sheep wool look a lot alike, too. In addition, Angora wool, like sheep's wool, can stretch. The two types of fur differ in that Angora wool is smoother,

### Ancient Angora

The origin of Angora goats dates back thousands of years to Turkey. In fact, the word *Angora* comes from the name of the ancient Turkish city Angora, which is present-day Ankara. The first Angora goats came to the United States in 1849. Today, about 1.2 million Angora goats are raised in the United States. According to the Mohair Council of America, Texas raises 90 percent of all U.S. Angora goats.

stronger, and often warmer than sheep's wool. The correct term for wool that comes from the Angora goat is *mohair*.

Mohair is not a soft wool, but rather a strong and durable one, since it comes from the coarser overcoat. Because of its durability, mohair is often used in products that get a lot of wear, such as furniture coverings and carpeting. Mohair is also used in clothing, slippers, blankets, wigs, paint rollers, and even toys.

Angora goats are considered the most efficient animals at making fiber. Perhaps that's because Angora goats are usually sheared not once, but twice a year. As with sheep, Angora goats are often shorn for the first time in late winter or early spring, right before kids are born. The kids nurse for the next few months. When they stop nursing, usually in July or August, the goats are shorn again.

## Cashmere

*Cashmere* is the fiber that comes not from one particular breed of goat, but from goats that typically live in cooler climates. The word *cashmere* derives from a mountainous area called Kashmir, which lies between Pakistan and India. Goats from this mountainous region are commonly called Kashmir, or cashmere, goats, although the word technically applies to the goats' soft fur, not to a specific goat breed.

*A model shows off the shaggy mohair dress that she wears under a matching gray striped fur coat at a fashion show in Milan, Italy, in March 1998.*

Unlike Angora goats, which are shorn twice a year, goats that produce cashmere are shorn only once a year. As the days begin to grow shorter—between June 21 and December 21—the goats' fur begins to grow. When the fur stops growing, the goats begin to shed. A few weeks before the shedding period is the time when most cashmere is harvested. The cashmere is shorn from the goat in a similar manner to how wool is shorn from a sheep. If the goats are not shorn before shedding, then the cashmere can be combed out as it is shed. Since the fur is falling out anyway, gently tugging the fur free and combing out the already-loose fur is not painful for the animal.

Cashmere is much softer than mohair, but it is much less plentiful. A typical adult Angora goat produces as much as 13 pounds (5.9 kilograms) of fiber a year. In comparison, an adult goat with cashmere down produces only about 0.5 pounds (0.23 kilograms) of fiber a year. Because of the scarcity of cashmere, clothing made with cashmere is expensive to produce, and therefore expensive to buy.

Even so, people enjoy the soft feel of cashmere. Products include sweaters, scarves, gloves, and even socks. Cashmere can be white, gray, black, or tan.

## CAMELS AND COMPANY

Camels belong to their own animal family, called, simply enough, the camel family. Animals that belong to the same animal family have similar characteristics. All members of the camel family have two wide toes on each foot. They also have longish faces, and their upper lips are split in the middle. Animals of the camel family are plant eaters, and they are ruminants. *Ruminants* are animals that have special stomachs that allow them to swallow food and then bring it back up again to be chewed more thoroughly. Most ruminants have stomachs with four parts. Animals of the camel family have stomachs with three parts. Members of the camel family also all

*A pair of Bactrian camels, originally from Mongolia, look across the pasture at the Tregellys Fiber Farm in Hawley, Massachusetts.*

have fur that can be used for clothing.

As with some goats, the undercoat of camels is very soft. Although camel hair may be used alone to make clothing, it is often blended with wool. Apparel made with camel fur usually has a soft, rich, brown color. Camel fiber is usually used to make camel-hair coats.

*Alpacas* and *llamas* are animals that come from the Andes Mountains of South America. *Guanacos* are considered the wild relatives of alpacas and llamas, which for the most part have been domesticated. In the United States, people have begun raising llamas and alpacas on farms for a variety of purposes, one of which is for their fur.

According to the Alpaca Owners and Breeders Association, alpaca wool is comparable to cashmere. The wool is as soft as

## Meet the Camel Family

Only a few animals belong to the camel family.

- dromedary camels (also called Arabian camels), which have one hump

- Bactrian camels, which have two humps

- alpacas

- guanacos

- llamas

- vicuñas

cashmere and comes in twenty-two natural colors—eighteen more than cashmere. The fiber is stronger and lighter than sheep's wool. In addition, the association explains that raising alpacas is easy and that the fur "is clipped from the animal without causing it injury."

The *vicuña* is a wild animal; it has not been domesticated like its alpaca and llama cousins. Living in the Andes Mountains of Ecuador, Peru, and Bolivia, it has a long wooly coat to protect it from the cold. Unfortunately, vicuña fur is highly prized, and the animals are often hunted for their fur. Efforts have been made to protect vicuñas, and therefore, vicuña fiber is very rare and expensive.

## SO WHAT'S THE PROBLEM?

Except for the vicuña, which is hunted for its fur, most of the animals from which people harvest clothing fibers appear to live safe and healthy lives. Why do some people object to these animals being raised and sheared for their fur?

One concern is the general health of the animals. Since fur is often removed before they shed, the animals risk the chance of being cold. Animal rights groups claim that this is inhumane. However, if properly cared for, the animals should not suffer from the weather. One source on raising goats even instructs caretakers to "make sure your sheared or combed goats have snug housing and plenty to eat."

Another reason animal rights groups object to harvesting animal fibers is that the animals are confined. They are not allowed to live out their lives in a natural setting. However, people who raise the animals quickly point out that these animals have already been domesticated—they have been removed from the wild and have become accustomed to living with people over generations.

## Llama Limits

Llama fur is coarser than alpaca fur. Although it is used as fiber for clothing, it is not as highly sought after as the softer alpaca fur. In South America, llamas are traditionally used as pack animals, carrying as much as 200 pounds (90 kilograms) as they are led by a handler. In the United States, some people have taken to using llamas as pack animals for hiking trips, strapping equipment to their backs. However, people do not saddle up and ride llamas.

## CHAPTER 7

# Leather

The part of an animal that has been discussed in the previous chapters is animal fur. The fur is a covering that grows from the animal's skin, just as human hair grows from the skin. But animal skin itself is also used to make clothing. That's what *leather* is—animal skin that has been specially treated so that it won't decay and then further treated so that it can be changed into a material from which a variety of clothing products are made. Such products include coats, jackets, vests, shoes, and pants. Leather can also be made into other items, like wallets, purses, belts, and furniture.

Depending on how the leather is processed, it can be hard, like leather from which shoes are made, or soft, like leather for gloves. The process of turning animal skins into leather is called *tanning*.

##  TANNING TRADITIONS

Tanning is not a new craft. In fact, historians believe that people thousands of years ago were familiar with the tanning process. A painting found in Egypt that dates back about 3,500 years depicts men working with animal hides.

Native Americans also tanned animal skins to make a variety of items, such as moccasins, clothing, and even tepees. Native Americans used the skins from animals that they had hunted, such as deer and buffalo, depending on where the natives lived. Their tanning process involved washing the hides, burying them until the fur became loose, and then scraping the fur away.

*This painting by Karl Bodmer, done in the mid-1800s, shows a native Plains warrior wearing tanned animal-skin clothing.*

## Manhattan Man

Peter Minuit was from the Netherlands. He had been sent to the New World by the Dutch West India Company. He settled on the island of what is today Manhattan, part of New York City. Minuit is the man who reportedly bought the island from the Native Americans in 1626 in exchange for a few trinkets worth about $24. The island was named New Netherland, and the settlement of New Amsterdam on its southern tip was its capital. It was here that Minuit built his tanning machine.

People had been making leather in Europe as well. When European settlers arrived in North America, they brought many leather products with them. However, as more settlers arrived, more leather products were needed. Some trading was done to obtain leather from the Native Americans, but the European settlers wanted leather that was a bit sturdier than the leather that the Native Americans offered. They began making their own leather from such animal skins as deer and elk, applying their own leather-making methods.

A man named Peter Minuit is responsible for the first machine built in North America to process leather. One of the ingredients that can be used in tanning comes from tree bark. The machine that Minuit built aided the process of grinding tree bark for use in tanning. From his one small tanning plant grew others, and today, the leather industry in the United States earns over $1 billion each year in the sale of leather goods.

##  THE TANNING PROCESS

Turning animal hide or skin into leather is not an easy or quick process. The ultimate goal of the tanning process is to ensure that the animal hide will not rot or shrink. It begins with the raw animal hide itself. This first step is called *curing*. During curing, the animal hide is dried, which is accomplished by covering the hide with salt. Sometimes, the salt is simply poured over the animal hide. In this case, the curing process takes up to thirty days. A much quicker way is called *brine-curing*. The hides are placed in a vat that contains water with large amounts of salt, or *brine*. Curing leather this way takes only about sixteen hours, so it is the most common method used.

Next, the hides are soaked in water to cleanse away the salt and any remaining blood and dirt. This stage can take several hours or several days. A machine then scrapes the inside of the hides to remove the animal flesh. The hides are soaked again for several days in a solution of water and limestone. Afterward, another machine scrapes the outsides of the hides to remove the animal fur or hair. Workers may also scrape both sides of the hides by hand to remove any extra flesh or hair.

The hides are soaked again, until finally the actual tanning process can begin. There are two types of tanning methods: *vegetable tanning* and *mineral tanning*.

Vegetable tanning is a much longer process that occurs over several weeks. Suspended on frames, the hides are swished through vats of tanning solutions. These solutions are made of *tannin*—a substance taken from tree bark, leaves, and fruit. The solutions prevent the hide from decaying and shrinking. The strength of the tannin solutions is gradually increased each week.

Leather produced from vegetable tanning is sturdy and strong and is used in such products as shoe soles, belts, and luggage.

Mineral tanning, also called *chrome tanning,* does not take as long to complete. The tanning solution this time is made from salt compound of chromium. The hides are placed in large, dryerlike machines that tumble them in the chromium-sulfate solution. This process takes less than a day to complete, because the solution penetrates the skin more quickly than the vegetable tanning solution. Other chemicals can also be used to produce different types of leather. Leather processed this way is usually softer than leather processed by vegetable tanning. Gloves, purses, the tops of shoes, and other clothes are usually made from mineral-tanned leather.

## WHERE DO THE ANIMAL HIDES COME FROM?

Long ago, when people traditionally hunted animals for food, animal hides were a by-product of hunting, as were most parts of animals. Today, animal hides are a by-product of the meat industry. Most leather is derived from skins taken from cows that have been killed for their meat. The same is true of leather made from the hides of calves, sheep, lambs, pigs, goats, and kids.

In fact, animal hides are the most important by-product of the meat industry. (Meat is the primary product.) According to the USDA's Livestock Marketing Information Center, in March 2001 the hide from a 1,000-pound (450-kilogram) steer sold for $60.86, which was 62 percent of the total amount of money earned for the steer's by-products.

Other animals, often referred to as "exotics," are raised specifically for their hides. These animals include alligators, snakes, and ostriches. However, most leather is made from animal skins that come from the meat industry.

*This herd of cows feeds together in close quarters on a factory farm in Europe.*

## ⌬ SO WHAT'S THE PROBLEM?

Many people rationalize that because the animals from which most leather comes are killed for their meat, it is okay to wear leather. It makes sense. After all, why should you throw away a cow's hide if the animal has already been killed for its meat? Discarding the hide seems like a waste. If you can eat a hamburger in good conscience, then wearing the skin of a cow in the form of a leather jacket should not be objectionable.

It is, however, to animal rights activists and animal welfare groups. Their objections lie mainly in the way that farm animals are raised for their meat. The problem for animal activists has to do with a modern-day practice called *factory farming*. Because of poor treatment of animals on factory farms, many animal activists object to eating meat raised on factory farms. (In fact, most animal rights groups object to eating animals at all.)

Here is an example of life on a factory farm. Imagine that you are an animal, living in a cramped, dark box, totally alone. You can barely move, and you never associate with other animals that are just like you. Or imagine that you are an animal in a huge room packed with thousands of other animals. You're so crowded that you barely have room to stretch or walk. This is your existence, day in and day out, with no change, no hope for anything different or better.

This is the life of many animals on factory farms. On pig farms, pigs are jammed into stifling sheds, and female pigs spend their lives in tiny crates, giving birth to litter after litter of piglets. Calves are placed in small stalls, which provide little room for movement. Other animals on factory farms, including dairy cows, beef cattle, chickens, and even geese and fish, also suffer from dark, cramped quarters in order to make room for as many animals as possible. Other practices on factory farms include branding, tail-docking, and castrating animals without anesthesia.

- *Branding* is the process of burning a mark onto an animal's skin as a way of identifying to whom the animal belongs.

- *Tail-docking* is the process of cutting off a portion of an animal's tail.

- *Castrating* is done to male animals. The animals' testicles— the body parts that produce sperm, which is necessary for reproducing—are removed, sometimes without anesthesia. (*Anesthesia* is the numbing of a body part before surgery; a substance that produces this loss of feeling is called an *anesthetic*.)

73

*This model's high-fashion look consists of a pony-skin coat and brown leather pants designed by Gianni Versace.*

To animal rights and animal welfare activists alike, such practices on factory farms are inhumane. Farmers argue that the welfare of their animals is of the utmost importance because a healthy animal will produce healthy products. Yet pictures and investigations have uncovered serious problems with animal care on some farms.

Animal rights groups offer another strike against the leather industry—pollution. They point out that all the chemicals used to turn animal hides into leather are harmful to the people who work with these chemicals. According to PETA, "More than 95 percent of leather produced in the U.S. is chrome-tanned. All wastes containing chromium are considered hazardous by the U.S. Environmental Protection Agency [EPA]." PETA claims that waste products that have the chemical chromium are dangerous, according to the EPA. The EPA is a government agency responsible for protecting the natural environment.

The leather industry claims that leather is "eco-friendly." Because leather comes from animals, leather products will eventually biodegrade when discarded. When something *biodegrades,* it is broken down naturally by bacteria and other microbes. Products made from artificial materials, like plastics, will not biodegrade when thrown away. "A good rule of thumb is, if you feel comfortable eating meat, you can feel comfortable wearing leather," states the Leather Apparel Association on its Web site.

## CHAPTER 8

# From the Sea

Throughout history, people have taken animals from their natural habitats and used them for clothing. Although most of these animals live on the land, some animals used for clothing come from the sea. The most popular marine animals hunted for clothing are seals, although other animals throughout the world, such as walruses and sharks, are also killed for clothing.

Native peoples of Canada and Alaska have traditionally depended on marine mammals for food, clothing, and other items. The Aleut of Alaska hunted the Steller sea lion, which provided not only food, but boat covers made from the animals' hides, tools made from the animals' bones, boot soles made from the animals' flippers, and clothing made from the animals' intestines. Today, although hunting marine mammals is against the law in some areas, Native Americans are often allowed to hunt them if it is a cultural tradition.

##  FUR SEALS

When one conjures up the image of a seal, one probably pictures an animal with a smooth, sleek hide. However, all seals and sea lions have some amount of fur on their bodies. Seals with the most obvious amount of fur are typically the *fur seals*.

Fur seals are actually considered part of the sea lion family. Sea lions differ from seals in that they have visible ears. Their flippers are also different. The front flippers of the seal are much smaller than the front flippers of a sea lion. Also, the hind

*These sea lion pups play near the shore of*
*San Cristobal Island in the Galapagos Archipelago.*

flippers of seals point backward; the hind flippers of sea lions point forward. A sea lion's flippers enable the animal to move more easily on land than a seal.

The male sea lion has a ring of fur around his neck, like a mane, which is how the animals got the name "sea lion." Fur seals tend to have thicker fur and much more of it. There are two types of fur seals: the South American fur seal, which lives in southern Pacific and Atlantic Ocean waters; and the Northern fur seal, which lives in the Bering Sea, between Alaska and Russia, and the Sea of Okhotsk, a body of water in Russia. Both the Bering Sea and the Sea of Okhotsk feed into the northern Pacific Ocean.

Fur seals venture on land mostly to mate and give birth. Male fur seals, or *bulls,* are very territorial and often fight each other for the best spots on the beach. Female fur seals, called *cows,* stay with their newborns for about a week, during which time the pup drinks the mother's milk. Afterward, the cow periodically returns to the water to find food. She then returns to her pup, calling to it and listening for the pup's cry. A pup stays with its mother for one to three years, or until the next pup is born.

## Meet the Pinnipeds

*Pinnipeds* are an order of animals made up of three animal families: sea lions and fur seals; walruses; and earless seals, or true seals. Pinnipeds are marine mammals that live a portion of their lives in the ocean but need to breathe air and spend some time on land. They have fur, and the females give birth to live babies and feed their babies milk from their bodies. The characteristics that make pinnipeds similar are flippers, the need to live part of their lives on the land, and the fact that they are *carnivores*—they eat meat.

*A seal lies on a piece of floating ice in the White Sea near Arkhangel'sk, Russia, in May 2001. A Russian shipping company sent a vessel to save hundreds of thousands of baby seals who were on the brink of starvation because high winds had pushed the ice floes into deeper waters and left the seals trapped in an area where most food was too deep for them to reach.*

Like many animals with fur, the fur seal was hunted to extremes. The fur of these seals was highly sought after, and hundreds of thousands were killed in the late 1800s and early 1900s. Between 1899 and 1909, about 600,000 Northern fur seals were killed, leaving a population behind of only between 200,000 and 300,000. Fur seals in the Southern Hemisphere suffered a similar fate, with even more drastic numbers killed. At one point in the early 1900s, it was estimated that only fifty Antarctic fur seals were left. Efforts were made to protect the animals, and today, both species have recovered to populations of about 1 million.

# @ HARP SEALS

*Harp seals* are not in danger of becoming extinct. Living in northern Atlantic and Arctic Ocean waters, they can be found in Russia, Greenland, and Canada. Those that live in Canada—in Newfoundland and the Gulf of St. Lawrence—and as far east as Greenland are called Northwest Atlantic harp seals. Researchers estimate the Northwest Atlantic harp seal population to be over 5 million.

Hunting harp seals in Canada is legal, although there are limits. Hunting can be conducted only from November 15 to May 15, but the times can change if requested by the sealing industry. Other seals that live in Canada, such as hooded seals, may also be hunted at this time. In addition, only a certain number of seals can be harvested per hunting season. Only 275,000 harp seals and 10,000 hooded seals may be killed. These numbers, established by the Department of Fisheries and Oceans (DFO) of Canada, are called the Total Allowable Catch (TAC).

A major controversy surrounding the killing of harp seals concerns not a decline in population—harp seals, for the moment, are plentiful—but the age at which many harp seals are killed and the ways in which they are slaughtered.

Harp seal pups are born in late February to early March. The mothers give birth on floes of ice, and they stay with and nurse their young for about two weeks. After that time, the harp seal pups are on their own. Since many pups are born in the same area, these new pups form family units, called *herds*. Unlike adult harp seals, young harp seals have a thick, white coat of fur from the time they are born until they begin to shed the fur, or *molt,* usually at about two weeks of age.

Once mother seals have left their pups, the pups are extremely vulnerable to both natural predators and human hunters. According to the animal welfare group International Fund for Animal Welfare, 80 percent of all harp seals killed are about twelve days to a year old. These young animals are

# The Marine Mammal Protection Act

In 1972, the United States passed the Marine Mammal Protection Act (MMPA). Its goal is to protect marine mammals that live in U.S. waters. Therefore, it is illegal not only to kill marine mammals in the United States, but to import marine mammals and products made from marine mammals, such as coats made from seal fur. In addition, the MMPA states that

@ certain species and population stocks of marine mammals are, or may be, in danger of extinction or depletion as a result of man's activities;

@ such species and population stocks should not be permitted to diminish beyond the point at which they cease to be a significant functioning element in the ecosystem of which they are a part, and, consistent with the major objective, they should not be permitted to diminish below their optimum sustainable population level;

@ measures should be taken immediately to replenish any species or population stock which has diminished below its optimum sustainable level;

@ there is inadequate knowledge of the ecology and population dynamics of such marine mammals and of the factors which bear upon their ability to reproduce themselves successfully; and

@ marine mammals have proven themselves to be resources of great international significance, aesthetic and recreational as well as economic.

What this means is that

@ some animals may be endangered because of the things that people do;

- the numbers of animals that are endangered should not be allowed to go below a certain level;

- people need to try to save these animals if their numbers do go below a certain level;

- people need to learn more about how sea mammals have babies; and

- mammals of the ocean are important to people around the world.

usually clubbed to death or killed with a *hakapik*—a hammerlike object with a large, deadly ice pick on one end. Older seals may be clubbed or shot to death.

However, according to the DFO and the Canadian Sealers Association (CSA), it is against the law to hunt harp seals that are so young that they still have their white fur. The CSA says that it has been illegal to hunt "whitecoats" since the 1980s, and hunters who do are charged by the DFO. In addition, the DFO specifically states in its Atlantic Seal Hunt 2001 Management Plan: "Whitecoats (harp seal pups) and bluebacks (young hooded seals) may not be hunted."

Answering allegations about the inhumane ways in which seals are killed, the DFO refers to Section 8 of Canada's Marine Mammal Regulations: "Persons can only dispatch marine mammals in a manner designed to do so quickly. Under these regulations, seals may be killed only by the use of high-powered rifles, shotguns firing slugs, clubs and hakapiks."

Further, even though the TAC for harp seals is 275,000, the numbers recorded by the DFO have been less. In 1999, 244,522 harp seals were harvested, and in 2000, the number was only 91,602. Although a lot lower than the 275,000 allowed, that's still 91,602 more than animal rights groups would like.

*In January 2001, young sea lions wait to be released by Galapagos National Park workers on Santa Fe Island in the Galapagos Archipelago.*

# THE SEAL PELT INDUSTRY

In Canada, the seal pelt industry is not quite the moneymaker that the fur industry is in the United States, yet it is still very valuable. The fur industry in the United States brings in over $1.5 billion. In contrast, the seal industry in Canada makes about $11 million, as reported by the CSA for 1996. Other products from seals include oil and meat. Leather made from sealskin is used to make a variety of products, including coats, jackets, hats, and briefcases. The CSA mentions that "fashionable fur products," seal fur coats designed by world-class fashion designers, are becoming popular in the international market.

But what about the United States? Will seal fur from Canadian harp seals ever be found on clothing in this country? "Potential markets in the U.S. remain closed due to restrictions in its Marine Mammal Protection Act," states the DFO. However, in its Atlantic Seal Hunt 2001 Management Plan, the DFO says,

> Sealing industry interests would like the United States to remove the prohibition on the import of seal products under its Marine Mammal Protection Act (MMPA). This prohibition has been in place since 1972 and the act is currently under review. The Department of Foreign Affairs [of Canada] has the lead on this issue [and] is presently developing a plan in an effort to open the U.S. market to sealing product opportunities.

This means that the DFO would like the U.S. government to allow seal products to be sold in the United States.

CHAPTER 9

# Illegal Entry

The MMPA makes it illegal to sell products made from marine mammals or bring them into this country. However, marine mammal products are not the only animal products that are against the law. It is also illegal to bring into the United States products made from any endangered animals.

In fact, many countries around the world ban the entry of such products. These countries work in conjunction with an organization called the Convention on International Trade in Endangered Species of Wild Fauna and Flora (CITES). The main goal of CITES is to prevent the trade of endangered animals and

## How Many Are Protected?

CITES lists approximately 5,000 animals and 25,000 plants as part of its no-trade program. The plants and animals are assigned to different categories, or appendixes, which explain their CITES status.

**Appendix 1**—These animals and plants are "threatened with extinction." Trade is allowed only in very special circumstances.

**Appendix 2**—These animals and plants are "not necessarily threatened with extinction," but trade must be controlled.

**Appendix 3**—These animals and plants are protected in at least one country.

plants between countries. Once animals or plants have been given endangered status by CITES, most of the countries that are part of CITES ban the trade of those animals, plants, and animal and plant products. To date, 146 countries belong to CITES, including the United States.

Even though certain animals are protected through CITES, illegal trade does occur. The elephant, snow leopard, and Tibetan antelope are often hunted and traded illegally for their fur and other body parts to be used as clothing or jewelry.

## @ TIBETAN ANTELOPE WOOL

One animal that has recently received a lot of attention due to its endangered status is the Tibetan antelope. A shy animal, the Tibetan antelope lives mostly in the higher elevations of the steppes of Tibet and China. Males and females usually mate in

*In July 2000, Kashmiri weavers protest with their spinning wheels against the ban imposed on shahtoosh wool. Weavers and tradespeople whose jobs depend on the sale of the soft wool of the endangered Tibetan antelope say that thousands of people will starve if the government does not reverse its ban.*

## Word Play

In Persian, shahtoosh means "king of wools." Shahtoosh shawls are also called *ring shawls*. Because Tibetan antelope fur is so fine, people claim that a large shawl can be pulled through a ring that you wear on your finger.

late fall or early winter, and young are born six months or so later, in May or June. Tibetan antelope aren't very big, measuring a little more than 3 feet (0.9 meters) tall at the shoulder and weighing between 55 and 77 pounds (25 and 35 kilograms). Tibetan antelope, also called *chiru,* have beautiful long horns that grow upward in two black, delicate-looking spears.

The Tibetan antelope isn't killed for its horns, however, but for its fur. Called *shahtoosh,* the fur is described as some of the finest hair in the world—softer and lighter than cashmere. The width of each hair fiber is three-quarters that of cashmere, and one-fifth that of human hair. The fur is most often woven into shawls by weavers in the Indian state of Jammu and Kashmir.

Unlike cashmere goats, however, which are shaved for their fur, Tibetan antelope are killed. In fact, it takes the fur of between three and five Tibetan antelope to make one shawl. The price of such a shawl is high—between $1,000 and $5,000. The shawls are considered luxury items and are highly prized.

Many organizations have become concerned for the survival of the Tibetan antelope. Researchers believe that several million of the animals lived in the early 1900s. However, by 1995, the Wildlife Conservation Society estimated that only 75,000 remained. China's State Forestry Administration estimates that about 20,000 Tibetan antelope are killed each year. The Tibetan antelope is listed as an Appendix 1 animal by CITES.

Even so, illegal hunting, or *poaching,* does occur. Officials believe that once the antelope are killed, the pelts are smuggled into Jammu and Kashmir, where trade in shahtoosh is actually

## Antelope Orphans

One of the tragedies of the illegal trade of Tibetan antelope wool is the orphans that are left behind. According to the International Fund for Animal Welfare, "Poachers gun down females in their calving grounds and leave the babies to die." People who try to stop poachers often find themselves caring for young antelope orphans.

legal due to cultural traditions. Here, shawls are woven and then either sold in India or smuggled into other countries. In the United States, several clothing dealers have stood trial for bringing shahtoosh shawls into the country. The guilty parties were fined, but not jailed. Animal rights groups had hoped for stiffer penalties.

##  ELEPHANT IVORY

Elephants are the largest animals to walk on land. Because they stand 11 feet (3.3 meters) tall and weigh between 8,000 and 13,000 pounds (3,600 and 5,850 kilograms), one would think that an elephant might be killed for its skin. After all, there certainly is plenty of it. This is not the case. Elephants are often hunted and killed for something much smaller in comparison to their overall body size—their tusks.

Elephant tusks are made of ivory, and ivory was once a hot commodity on the international market. An elephant tusk is like an elongated tooth made out of a hard, off-white substance. Craftspeople carve the ivory for a variety of products, including buttons and decoration for clothing, jewelry, statues, and piano keys. According to the World Wildlife Fund, ivory demands skyrocketed in the 1970s, leading to the slaughter of many elephants.

Because of CITES, it is illegal to trade ivory in international markets in most countries around the world. CITES has established a project aimed specifically at observing elephant populations: Monitoring the Illegal Killing of Elephants (MIKE). The project's benefits include the following:

> For the first time, at an international level and on a consistent scientific basis, MIKE will assess, in the selected MIKE sites, the levels and trends of elephant populations and illegal killing. MIKE will attempt to identify the reasons for any change in population trends, including whether any changes can be attributed to CITES decisions to allow or suspend trade in elephant products.

This means that people working on the MIKE project will look closely at how many elephants are alive and how many are killed. The group will also try to find reasons why the numbers of elephants change and if the selling of products made from elephants is one of those reasons.

*In May 1997, George Gadzai of the Zimbabwe Parks Board poses inside the state ivory warehouse in Harare, Zimbabwe. The warehouse holds more than 3,000 elephant tusks.*

# SNOW LEOPARD FUR

Snow leopards are beautiful members of the cat family. They are not overly large, measuring about 3 to 4 feet (0.9 to 1.2 meters) long. Females weigh about 80 pounds (36 kilograms), and males weigh 100 to 120 pounds (45 to 54 kilograms). For comparison, a male lion can weigh up to 530 pounds (239 kilograms), and a male tiger can weigh up to 600 pounds (270 kilograms).

Living in the cold, mountainous regions of Russia, China, Nepal, Pakistan, and other countries of Asia, snow leopards are hard to find. But that doesn't stop people from hunting them, nor does their status as an endangered animal. Snow leopards have a beautiful pelt—long white hairs with dark, distinctive spots. In addition to being hunted for their fur, snow leopards are also in danger because of loss of habitat.

Many organizations list the snow leopard as endangered. The International Union for Conservation of Nature and Natural

*A pair of male snow leopard cubs are seen at the San Antonio Zoo in San Antonio, Texas. The endangered cubs spent nine weeks in the nest area with their mother and then became part of the zoo's snow leopard exhibit.*

# The Tip of the Iceberg

The animals described in this chapter are only a few of those threatened by illegal hunting and trading for the clothing industry. Others are listed below.

- Although usually killed illegally for traditional Chinese medicines, tigers are also killed for their fur.

- The hawksbill sea turtle is a marine animal that has been hunted for a variety of reasons, including turning the shells into jewelry and the hides into leather.

- Although coral might seem to be more rock than living being, it is an animal, and in many countries, it is illegal to remove coral from the ocean. If a country deems that it is illegal to harvest coral, then it is illegal to bring coral from that country into the United States.

Resources calls the snow leopard "critically endangered," and the United States Endangered Species Act lists the snow leopard as "endangered." CITES lists the snow leopard as an Appendix 1 animal. Researchers estimate that only a few thousand snow leopards still exist.

## HOW WILL THE GOVERNMENT KNOW?

Imagine that you have traveled to a foreign country. In a gift shop, you find some unique souvenirs. One is a very beautiful, very soft shawl—a perfect gift for your mother or grandmother. You also find distinctive earrings made of ivory. In another shop, you see a belt covered with an interesting spotted fur, or perhaps a wallet made from a soft leather. You buy all these

## Mission Statement

The primary mission of the National Fish and Wildlife Forensics Laboratory, as stated on its Web site, is to:

- identify the species or subspecies of pieces, parts, or products of an animal

- determine the cause of death of an animal

- help wildlife officers determine if a violation of law has occurred

- identify and compare physical evidence in an attempt to link suspect, victim, and crime scene

That is, the mission of the National Fish and Wildlife Forensics Laboratory is to figure out from which animal an animal part or product comes; how the animal died; if a crime was committed by killing the animal; and who killed the animal and where.

items to take home, packed along with the rest of your clothing, not giving them a second thought.

Upon entering the United States, however, you get stopped randomly for a customs inspection, and all the items you've bought are confiscated. Why? And what did you do wrong?

You have brought into the country items that are suspected of being from endangered animals. We use the word "suspected" because tests need to be conducted in order to determine the materials from which the items were made.

In most cases, these suspected items are sent to a government agency—the National Fish and Wildlife Forensics Laboratory. Here, scientists examine items that are suspected of being made from endangered animals. The scientists look at a variety of things: blood samples, tissues samples, bones, teeth, claws, tusks, hair, hides, fur, feathers, and leather. If it is determined that an item was indeed made from an endangered animal, watch out! Bringing items into the United States that are made from endangered animals, as listed by CITES, can lead to hefty fines.

Endangered animals are still being killed illegally for many reasons, including being made into clothing. Only when people stop buying these products will poachers realize that the animals are more valuable alive than dead.

*This Kemp's Ridley sea turtle, weighing 80 to 100 pounds (36 to 45 kilograms), heads back to the water after getting a satellite transmitter attached to her back so that her movements can be tracked.*

## CHAPTER 10

# From Fiber to Fabric

The clothing, or *apparel,* industry is big business. In 1999, sales of apparel—which includes all clothing, including shoes—in the United States were over $170 billion. That figure was up 4 percent from the previous year. The table below breaks down the sales according to men's, women's, and children's clothing.

## Clothing Sales

| | |
|---|---|
| Girls' | $9,932,000,000 |
| Boys' | $13,515,000,000 |
| Men's | $56,514,000,000 |
| Women's | $96,006,000,000 |

Let's face it—people need clothes. We need clothes to protect us from the weather. We also need clothes so that we don't walk around bare in public. We are the only animal on Earth that feels the need to cover up.

We also wear clothes to communicate. Personal choices in clothing reflect our feelings and interests. Some people like to make bold statements, wearing bright colors and outlandish styles. Others wear more conservative clothing, with muted

*This young woman tries on a long sweater coat. Some parents, outraged by what they see on store shelves, say that they will not buckle under to pressures from the apparel industry to transform their daughters into clones of popular rock stars.*

colors and more basic fashions. Traditional clothing worn in countries around the world can identify cultural differences. Uniforms identify the jobs that people have or associations to which they belong, such as clubs, the military, or schools. With all these different reasons for wearing different types of clothing, materials for making clothing are essential.

##  THE TEXTILE INDUSTRY

Long ago, people made clothing from plant and animal products. Today, we still use animals and plants, or natural materials, for clothing, plus a variety of manufactured materials. Fabrics that are woven using plant fibers, animal fibers, or manufactured fibers are called *textiles*. The manufacturing of materials from fibers is the *textile industry*.

Processing any fibers into fabric follows the same basic process as turning wool into fabric. The fibers are first spun into yarn, and the strands of yarn are woven into fabrics. The fabrics are then dyed certain colors, or they may be printed with patterns or pictures. After a few finishing touches, the fabrics can then be made into clothing. Preparing the fibers and turning them into fabric is typically done at a textile mill.

Textiles are more than just clothing. Today, they include any products for which fibers have been spun and woven together. These products include ropes, nets, furniture coverings, rugs, canvases, bags, and many others.

## SILK FIBERS

Fibers that come from sheep and goats can be spun and woven into fabrics. These fibers are considered *animal fibers*. Another type of animal fiber is *silk*. Silk comes not from the fleece of a mammal, but from an insect—the silkworm moth.

*Tu Chuan-ming picks through silkworms in November 2001 as he ponders reasons for abandoning the family's silk business on the island of Taiwan. He must compete with silkworm farmers in China, who can sell their cocoons at much cheaper prices.*

Moths are similar to butterflies in a number of ways. They are both insects, with six legs and three sections to their bodies. They both have wings, along with antennae on the tops of their heads. Butterflies and moths both go through a series of changes called *metamorphosis*. It is this series of changes, from young insect to adult insect, that is vital to producing silk.

Moths begin life not as six-legged insects with wings, but as caterpillars. This is the first stage of the moth's life, called the *larva stage*. The caterpillars hatch from eggs laid by an adult female moth. Also called *silkworms,* the caterpillars eat the leaves of mulberry trees. As the silkworms grow, they shed, or *molt,* their skins. They do this four times, after which the silkworms are ready for the next phase of their lives—the *pupa stage.*

*Each of these fuzzy white cocoons created by silkworms is made of one continuous silk fiber that may stretch half a mile (0.8 kilometers) in length.*

# Insects 101

Although moths and butterflies are similar in some ways, they differ in several other ways.

- Butterflies are active during the day; moths are active at night.

- Butterflies' antennae are thin; moths' antennae are feathery.

- Butterflies rest with their wings held above their backs; moths rest with their wings spread out.

During the pupa stage, the worms hang from branches and spin cocoons around themselves. On a silk farm, the worms are placed in wooden frames to complete the pupa process. The material that the silkworms produce to make the cocoons is silk. It will take each silkworm about two days to spin its cocoon, spinning over 0.5 miles (0.8 kilometers) of unbroken silk fiber. Inside the cocoon, the silkworm begins to change, and eventually, it will emerge as an adult silkworm moth.

However, most silkworms on silk farms never reach adulthood. When the adult moths emerge from their cocoons, they break the silk fiber. But silk farmers want this fiber in one long strand. To obtain the fiber, the farmers place the cocoons into hot ovens, and the silkworm pupae die inside their cocoons. The thread of each cocoon is then carefully unraveled at a special factory, and eventually, the silk is made into silk fabric.

To animal rights groups, the process of making silk is unacceptable. Animal rights groups protect all animals, not just the "cute and furry" kinds, and silkworms are animals, too. According to PETA: "Consumers are beginning to question the unnecessary killing of even tiny, sometimes complex, and certainly feeling, creatures like these."

*Two combines pick cotton in a field in Scott, Arkansas.*
*The cotton boll fibers are plainly visible on these mature plants.*

# ◉ PLANT FIBERS

Not all fibers come from animals. Some fibers come from plants, also called *vegetable fibers*. They include cotton and flax.

## Cotton

*Cotton* is a plant fiber that comes from the cotton plant. In fact, cotton is the most popular form of textile material in the world—half of all textiles are made from cotton fibers.

Cotton is grown in fields and harvested in a manner comparable to food crops. On the cotton plant, the flower bud opens to reveal a white, fluffy ball of fiber, called the *cotton boll*. These bolls are picked either by hand or by a machine. They are then run through a *cotton gin*—a machine that separates the cotton fibers from the seeds to which the fibers are attached. The fibers are then sent to a textile mill, where they are spun into yarns and woven into fabrics.

Cotton is a strong fiber, and therefore, clothes made from cotton are usually sturdy and long-lasting. Cotton is also comfortable to wear, because it is usually soft against the skin.

## Flax

*Flax* is a plant fiber that comes from the stem of the flax plant. To harvest flax, the plant is plucked from the ground, roots and all, and left in the sun to dry, as well as to weaken the plant material that covers the stem. When weakened, the covering can be crushed, revealing the flax fibers inside.

Flax is used to create a type of material called *linen*. Linen fabric is made into a variety of clothing, such as shirts, dresses, and pants, as well as into tablecloths and napkins. People like to wear linen because it is a cool, comfortable fabric. However, linen fibers break easily, which is why linen clothing often has a lot of wrinkles.

# MANUFACTURED FIBERS

Manufactured fibers are fibers made not from the raw materials of plants or animals, but from other substances. There are two types of manufactured fibers. *Cellulosic fibers* are made from a chemical that comes from wood, called *cellulose*. Rayon and acetate are two fibers made from cellulose. *Synthetic fibers* are made from petroleum or natural gas. Synthetic fibers include polyester, nylon, acrylic, and spandex.

Manufactured fibers begin not as strands but as hot liquids. The liquids pass through a special machine, which molds them into fibrous strands. When one of these liquids cools, it hardens and is rolled onto spools, from which the fiber can now be woven into fabric. Here are some qualities of the most common types of manufactured fibers.

*This outfit—metallic nylon jacket over a heather-gray cotton/Lycra T-shirt tucked into light khaki cotton/nylon sailing shorts—demonstrates fabrics made from a combination of natural and synthetic fibers.*

- *Rayon* comes from cellulose. It can be made smooth and soft, like silk, but with similar properties to cotton, such as strong fibers.

- *Polyester* is a strong and versatile fiber. Polyester fibers are often mixed with natural fibers like cotton to make the overall material more durable. Polyester also doesn't wrinkle, which many people see as an advantage.

- *Nylon* is created from chemicals found in oil. It is stronger yet lighter than most other fibers. During World War II (1939–1945), nylon was not available to the public; it was used to make such war materials as parachutes, tires, and uniforms. Today, nylon is often associated with stockings.

- *Spandex* is a very stretchy material that returns to its original size or shape after being stretched. Spandex usually holds up well when washed.

## NO MORE ANIMAL FIBERS?

People first began to manufacture artificial fibers in the late 1800s. Today, a large variety of manufactured fibers can be used to make all types of clothing. Still, many people prefer animal fibers. They like the softness of cashmere and the warmth of wool. They enjoy the smoothness of silk and the durability of leather. Many also luxuriate in the feel of real fur.

Animal rights groups question the need for clothing made from animal products. These groups argue that with the ability to make clothing from plants and manufactured materials, the need to use animals for clothing has passed. Many others, however, disagree.

**CHAPTER 11**

# Debating the Issue

Animals are used in varying degrees to make clothing. Some animals, like some sheep and goats, are kept alive. Their fur is shorn from them, and it grows back to be shorn again. Some animals are killed for their meat, and their hides are a by-product. Cows, pigs, and some sheep are raised and killed for food, and then their hides are turned into leather. Other animals are raised on farms specifically to be killed for their furs, such as mink, while still others are hunted in the wild and killed for their furs, such as foxes and seals.

Obviously, fur-bearing animals are handled differently from domesticated animals. Yet is there a difference, ethically, between using animals for clothing and using them for food or companionship? Is it okay to cut wool from a live sheep but not okay to kill an animal for its fur? Is it okay for other countries to kill fur-bearing animals that are plentiful, such as seals, but not okay to kill other plentiful animals for fur, such as cats and dogs? If so, why? Animal rights groups see no difference.

## ANIMAL RIGHTS

Some people feel that animals should be allowed the same rights as people. This means that they should be allowed to live their lives in their natural environments without exploitation or interference from humans.

For example, PETA states on its Web site:

Animal rights means that animals deserve certain kinds of consideration—consideration of what is in their own best interests regardless of whether they are cute, useful to humans, or an endangered species and regardless of whether any human cares about them at all.... It means recognizing that animals are not ours to use—for food, clothing, entertainment, or experimentation.

What PETA is saying is that all animals should be thought about and given rights, no matter how big or small, how cute or ugly. PETA does not want people to use animals for any purpose—not for food, clothing, entertainment, or research.

The Animal Legal Defense Fund (ALDF) is a group that is trying to pass laws that recognize basic animal rights. ALDF has a petition on its Web site, and it encourages others to sign the petition and return it to ALDF. Called the Animal Bill of Rights, it states:

I, the undersigned American citizen, believe that animals, like all sentient beings, are entitled to basic legal rights in our society. Deprived of legal protection, animals are defenseless against exploitation and abuse by humans. As no such rights now exist, I urge you to pass legislation in support of the following basic rights for animals:

- The right of animals to be free from exploitation, cruelty, neglect, and abuse

- The right of laboratory animals not to be used in cruel or unnecessary experiments

- The right of farm animals to an environment that satisfies their basic physical and psychological needs

- The right of companion animals to a healthy diet, protective shelter, and adequate medical care

- The right of wildlife to a natural habitat, ecologically sufficient to a normal existence and a self-sustaining species population

*Dressed in furs, two PETA members sit in cages outside a department store in East Grand Rapids, Michigan, in December 1997. They are demonstrating to bring attention to what they say are extreme weather conditions and overcrowding on fur farms.*

- The right of animals to have their interests represented in court and safeguarded by the law of the land

The ALDF is saying that animals should be viewed as creatures with feelings and emotions, and accordingly, the animals should have certain rights. In simpler terms, these rights are as follows:

- Animals should not to be neglected or treated badly.
- Animals used in labs should not be treated cruelly or used in experiments that are not really needed.
- Farm animals should live as stress-free and naturally as possible.
- Pets *(companion animals)* should be well fed, housed, and cared for.
- Wild animals should be able to live naturally in their habitats.
- Animals should be able to have someone speak up for them in court if their rights are violated.

In some ways, the idea of animal rights sounds logical. Animals are, after all, living creatures, just as humans are. Like humans, animals do experience pain and some emotions. Also, no one could really argue with many of the points listed in the ALDF's Animal Bill of Rights. We would like our companion animals (cats, dogs, birds, and so on) to eat well, live comfortably, and receive medical treatment when needed. Animals in general should not be treated cruelly, abused, or neglected.

An Animal Bill of Rights is not exactly the same as "animal rights" as far as animal rights groups are concerned. In the ALDF's Animal Bill of Rights, the group allows animals to be used in some ways, as long as the animals are not treated cruelly. Animal rights groups, on the other hand, believe that no animals should ever be used.

If trapping is considered cruel, is it against the Animal Bill of Rights? What about silk farms? Do silkworms fall under the protection of the Animal Bill of Rights? If so, is it cruel to place silkworm cocoons in ovens to kill the pupae?

What about the word "exploitation"? To *exploit* something means to use it selfishly, without thinking how it is being affected. Are sheep being exploited when people use their fur for wool? Are cows being exploited when their skin is used for leather, even though they've already been slaughtered for meat? Are animals raised on fur farms being exploited for their furs?

"We can't stop all suffering," says PETA, "but that doesn't mean we shouldn't stop any. In today's world of virtually unlimited choices, there are usually 'kinder, gentler' ways for most of us to feed, clothe, entertain, and educate ourselves than by killing animals."

## ANIMAL RIGHTS VERSUS ANIMAL WELFARE

Some people feel that the principles put forth by animal rights groups are much too radical. FCUSA debates the issue of animal rights versus animal welfare on its Web site:

> The animal welfare philosophy is fundamentally different from the animal rights philosophy, since it endorses the responsible use of animals to satisfy certain human needs. These range from companionship and sport, to uses which involve the taking of life, such as for food, clothing and medical research. Animal welfare means ensuring that all animals used by humans have their basic needs fulfilled in terms of food, shelter and health, and that they experience no unnecessary suffering in providing for human needs.

The FCUSA points out that animal welfare permits the use of animals by people, which is not the same as animal rights. Groups like FCUSA and the National Animal Interest Alliance

(NAIA) believe that it is okay to use animals to fulfill human needs, as long as the animals are treated properly. As stated on NAIA's Web site: "Animal welfare requires humane treatment of animals on farms and ranches, in circuses and rodeos, and in homes, kennels, catteries, laboratories, and wherever else animals are kept."

The NAIA also says: "Animal welfare grows and improves as we learn more and more about animals, their behavior, and their management. Animal rights remains stagnant with its dogma of 'no more animal use ever.'" (*Dogma* is a principle or belief considered to be absolutely true by a particular group.) The NAIA believes that animal rights activists don't want people to use animals ever, even if using them means that we could learn more about them. However, the NAIA feels that only through learning about animals and by using them can animal welfare become better.

##  DIFFERENCES OF OPINION

People who raise, trap, hunt, and use animals for clothing see the use of animals differently than do animal rights groups and many animal welfare supporters. For example, fur trappers believe that they are helping the environment by trapping animals for fur, because they are controlling animal populations, as well as using a renewable resource. They say that it takes more energy to produce manufactured fibers than it does to produce clothing made from animal fur. In addition, they point out, clothes made from animals are *biodegradable*—when people are finished wearing them, they will break down and become part of the earth again. Clothing made from manufactured fibers will not. Therefore, animal fur is better ecologically.

Animal rights groups, on the other hand, argue that if the animals were allowed to live naturally, their populations would take care of themselves. Since people have the ability to make fibers from manufactured materials, they should do so and stop

*A worker at the Natural Fibers plant in Ogallala, Nebraska, handles the company-developed pillow and comforter stuffing made of goose down and milkweed fibers in July 2001. The company sold $1.4 million worth of pillows and comforters stuffed with the down-milkweed mixture in 2000.*

contributing to the suffering of animals. Animal welfare groups also agree that many trapping procedures are not in the best interests of animals and in fact, are often quite cruel.

Another example of how animal rights and welfare groups see things differently from people in the clothing industry is in the use of goose and duck down. Coats, vests, jackets, and quilts can be stuffed with *down*—the soft underfeathers of ducks and geese—to make them warmer. Saying "Down on down," PETA describes the down-obtaining process, mostly in Europe, as inhumane: "Plucking the geese causes them considerable pain and distress. Four or five times in their lives, they will squirm as a plucker tears out five ounces [140 grams] of their feathers."

However, according to many industry sources, the birds are usually first killed for their meat, and the feathers are removed afterward. So like cowhide, goose and duck down are by-products of the food industry. As the Food Safety and Inspection Service of the USDA says,

> When these birds are slaughtered, they are first stunned electrically. After their throats are cut (by hand, for geese) and the birds are bled, they are scalded to facilitate removal of large feathers. To remove fine pinfeathers, the birds are dipped in paraffin wax. Down and feathers, a very valuable by-product of the duck and goose industry, are sorted at another facility.

Perhaps the description of the way ducks and geese are killed for food is objectionable to you. Remember, choosing to wear clothing with duck or goose down is your decision to make.

## @ A PERSONAL DECISION

Some ways in which animals are used for clothing may be more difficult to accept than others. For example, when wearing a wool sweater, one can rationalize that the sheep was not killed, although animal activists would point out the possible harm inflicted on the sheep. When buying shoes made of leather, one could argue that the cow was killed for its meat anyway, so why not make use of the cow's hide?

On the flip side, the capturing of an animal in a leghold trap from which it struggles to free itself may seem barbaric. Raising animals for fur may seem inhumane, too. Even though the animals would not have been born if not for the fur market, the thought of these cute, furry creatures confined to cages for their entire lives may seem intolerable.

The ASPCA has this to say about using animals for clothing, specifically fur-bearers: "The ASPCA is totally opposed to the use of animal furs in clothing and accessories. No distinction is made between wild, caught and ranch-or-farm raised animals because there is tremendous cruelty involved in fur ranching and the slaughter of fur-bearers." Yet, according to animal activists, if you feel for the mink, then you must feel for the silkworm, too.

Coming to terms with animal rights versus animal welfare is a personal decision. This decision can be made only by researching the ways in which animals are used for clothing and seeing the problems, if any, for yourself. For example, a visit to a sheep ranch might dispel any concerns about how the animals are handled during shearing time—or it may arouse new ones. Seeing photos of animals caught in leghold traps might sway your opinion one way or the other. Visiting the Web sites of clothing industry associations, as well as animal rights and animal welfare groups, might give you a clearer picture of the debate.

Long ago, humans didn't have a choice about what they could wear. Clothing made from animal products was the norm. In today's modern world, people have many choices. They can choose to wear clothing made from animal fibers, plant fibers, or manufactured fibers.

Animal rights activists feel that no animals should be used for clothing. Animal welfare groups feel it is okay to use animals for clothing, as long as the animals are treated humanely.

Which position do you support? The choice is yours.

# Glossary

**cashmere**  wool that comes from goats that live in the region of Kashmir

**cotton**  a plant from which clothing material is made

**endangered species**  a species that could become extinct because existing populations are very small

**euthanasia**  the act of painlessly killing an animal

**extinct**  dead; having died out; refers to a species or population, not an individual organism

**fiber**  something that is used to make material or fabric for clothing; fibers can be made from animals, plants, or artificial materials

**flax**  a plant from which the cloth linen is made

**leather**  animal skin that has been specially treated so that it won't decay

**mohair**  wool that comes from the Angora goat

**pelt**  an animal skin with the fur still attached

**poaching**  illegal hunting and killing of animals

**shahtoosh**  the fur of the Tibetan antelope

**tanning**  the process of turning animal skin into leather

**textiles**  fabrics that are woven using plant fibers, animal fibers, or manufactured fibers

# Bibliography

 **BOOKS**

Dineen, Jacqueline. *Wool.* Hillside, NJ: Enslow Publishers, 1988.

Hollenbeck, Kathleen M. *Islands of Ice: The Story of a Harp Seal.* Norwalk, CT: Soundprints Division of Trudy Communications, 2001.

Miller, Sara Swan. *Goats.* New York: Children's Press, 2000.

Perl, Lila. *From Top Hats to Baseball Caps, From Bustles to Blue Jeans: Why We Dress the Way We Do.* New York: Clarion Books, 1990.

 **WEB SITES**

**The American Society for the Prevention of Cruelty to Animals (ASPCA)**
www.aspca.org

**The Canadian Sealers Association (CSA)**
www.sealers.nf.ca

**Fabrics.net**
www.fabrics.net

**Fur Commission USA (FCUSA)**
www.furcommission.com

**The Humane Society of the United States (HSUS)**
www.hsus.org

**National Animal Interest Alliance (NAIA)**
www.naiaonline.org

**National Trappers Association (NTA)**
www.nationaltrappers.com

**People for the Ethical Treatment of Animals (PETA)**
www.peta-online.org

**United States Department of Agriculture (USDA)**
www.usda.gov

#  SOURCES

www.furcommission.com/farming/humane.htm

www.hsus.org/current/death_chart.html

www.conservewildlife.org/welfare3.html

www.leatherassociation.com/FFaq.html

www.dfo-mpo.gc.ca/seal-phoque/reports/
Mgtplan2001/sealplan2001_e.htm

www.cites.org/eng/programme/MIKE/part_I/MIKE.shtml

www.lab.fws.gov/mission.htm

www.animaljustice.org/petition/page1.asp

www.peta-online.org/fp/faq.html

www.furcommission.com/debate/index.html

# Index

*Note: Page numbers in italics indicate illustrations and captions.*